FORWARD

In 1953 Ward H. Cook, MD wrote *The Judging of Dahlias in Shows*; it was the first U.S. manual for the judging of dahlias. Since that time judges have referred to the original and all revisions as simply Cook's Manual. Dr. Cook's guide was one of the first publications to detail the evaluation of specific and important characteristics of the dahlia. Judges compared the cultivars they evaluated to the ideal or perceived perfect dahlia using points of reference that were listed on a scorecard for each characteristic. Through the years, judges have changed some of their methods and ideas about the dahlia. The revised manuals of 1968, 1983, 1989, 2002, and 2022 reflect transitions in dahlia thought and judging practice.

In 2020, American Dahlia Society (ADS) President Brad Freeman requested a committee be formed representing various regions of the ADS to make the judging manual current. This completed manual, representing the thoughts and suggestions of the committee members, has been reviewed and approved by the current Executive Board of the ADS.

The ADS is indebted to Ward H. Cook, MD for his original work and to the various committees that have contributed to all revisions.

The revision committee for this edition include:

Tony Evangelista
Brad Freeman
Art Haskins
Jo Jepsen
Marian Landerholm
John Morton
Lou Paradise
Karen Sanchez
Bob Schroeder
Terry Schroeder
Roger Walker
Marilyn Walton

Colin Walker, Design & Photography
Terry Schroeder, Secretary

Bulletin of the American Dahlia Society Inc. ISSN 0002-8150 – Volume 109, No 3. Published 4 times yearly.
Periodicals Postage Paid at Burbank, IL and additional mailing offices.
POSTMASTER: Please send address changes to
Bulletin-ADS attn: Elva Sellens, 2651 NW Loma Vista Drive, Roseburg, OR 97471

AMERICAN DAHLIA SOCIETY

Guide to Judging Dahlias

TABLE OF CONTENTS

CHAPTER 1

INTRODUCTION TO JUDGING DAHLIAS

Welcome. The evolution from a dahlia grower and exhibitor to a dahlia judge can be an interesting and rewarding one. This guide will act as your resource as you move through the many facets of dahlia judging. The knowledge you gain working through these chapters will give you a solid foundation to build your confidence about judging dahlias and to work with other judges.

The Judging of Dahlias in Shows (Cook's Manual) continues to evolve as exhibitors, judges, and dahlias grow and change. The ADS intends to update this manual as needed to ensure that an effective and contemporary document is available to dahlia judges and exhibitors. All exhibitors have the right to know the standards by which their entries will be judged and the information which competent dahlia judges employ when they determine the merits and faults of entries.

This revision continues to rely on the most current *Classification and Handbook of Dahlias* (CHD) for revising and/or adding technical definitions, exhibitor abbreviations, regional classification supplements, and classifications of cultivars according to form, size, and color. The ADS publishes this classification book annually.

Although judges will come to a dahlia show with knowledge of the recommended standards, all judges will view the exhibits from their own perspectives. Even though judges study these standards and use them as a measurable guide for making decisions, they will often not see an exhibit in the same way. Much of judging measures a matter of degree rather than right or wrong. In fact, the ADS encourages judges to arrive at their own decisions based solely on the merit of the entry. Judges should never succumb to selecting a winning entry because of personal prejudice or intimidation from another judge. They must be honest with and true to themselves in every decision they make.

Individuals who choose to judge accept certain responsibilities:

- ★ They will inform themselves of as many different cultivars as possible.
- ★ They will grow quality dahlias, learning superior horticultural practices.
- ★ They will observe and learn from competent judges who will serve as role models.
- ★ They will avoid outdated practices by continually upgrading their knowledge and experience.
- ★ They will convey a gentle, honest, and confident demeanor.

To meet requirements for promotion:

- ★ Judges must continue to grow and show dahlias.
- ★ Judges must attend dahlia shows.
- ★ Judges will work as team members with experienced judges to gain practical experience.
- ★ Judges should attend workshops and judging classes to advance their skills.

No two shows will ever have the same conditions, and the judge can garner insights that will significantly augment book knowledge. The knowledgeable and experienced judge has learned from training, experience, observation, and background information to recognize true beauty and perfection and to identify it without additional assistance.

The rules of good conduct define the ethics of judging. Ethical conduct is based upon honesty, empathy, integrity, sincerity, courtesy, intelligence, and wisdom. Judges have a particular duty to conscientiously follow the principles of ethical behavior.

TEXTS AND MATERIALS

Dahlia judges should possess a current CHD, current *Composite Listing of Dahlias*, the latest revision of *Guide to Judging Dahlias*, an ADS-approved color chart, a calculator, and a measuring device. They should have a keen understanding of the latest ADS scorecards and understand how to use them for both fully double and open-centered cultivars.

These materials will allow competent judges to record the physical attributes of the dahlia, determine the size and depth of a bloom, determine stem length, and bloom color. Through the application of these resources, judges must demonstrate that they know one dahlia form from another, that they can identify cultivar form based on the shape of a floret, and that they can determine the closest form of a cultivar.

QUALIFICATIONS OF A DAHLIA JUDGE

Judges must know a variety of cultivars and must be able to classify them according to form, color, and size; they must be up-to-date on the current classification descriptions. They must be able to recognize pest problems, diseases, and nutrient deficiencies. They must be rule conscious and sensitive to classification errors and misplaced entries. They should be able to judge with others who have varying temperaments and experience and they should be able to judge systematically and as quickly as possible.

Judges should:

- ★ Take refresher judging courses as required by the ADS.
- ★ Review and study the textual materials yearly, including: the most current edition of the *Guide to Judging Dahlias*, CHD, and the ADS seedling evaluation scorecards with the intent to master the information.
- ★ Always come prepared when judging dahlias. Bring judging tools and supplies: pencil, pen, small tape measure, *Guide to Judging Dahlias*, CHD, local show schedule, and if there is a possibility of judging seedlings, an ADS Color Guide, extra ADS Seedling Evaluation Scorecards (scorecards), and a small calculator.
- ★ Understand the relationship between a show schedule and the show's clerk sheets.
- ★ Understand the horticultural practices of dahlia growing.
- ★ Have a keen knowledge of many cultivars.
- ★ Possess willingness to travel and judge in dahlia shows.

SUMMARY

All judges:

- ★ Should continue to learn and grow, always aspiring to increase his/her awareness of what it takes to grow a good dahlia, to become a competent judge, and to be a genuine leader.
- ★ Should grow and show quality blooms revealing a competence and understanding of what it takes to bring a superior exhibit to a show.

- ★ Should continue to attend judging classes, dahlia workshops, and seedling seminars when offered – and should grow some seedlings themselves if possible.
- ★ Should try to enter dahlias in as many classes and sections as practical.
- ★ Must grow and show dahlias. Those judges who don't will have difficulty understanding the new forms and changes in definitions, which is required to be a competent judge.
- ★ Must remember that rules of judging are evolving over time. Gaining knowledge and experience is essential. There is no substitute for growing and showing to keep judges aware and up-to-date. A judge who no longer grows and shows dahlias will have difficulty providing leadership for his/her team.
- ★ Should encourage team members to use reasonable speed in arriving at a decision. Encourage less experienced judges to come to a final decision first.
- ★ Must remember that any collusion between judges indicating favoritism to or prejudice against an exhibitor is not appropriate and will not be tolerated.
- ★ Must remember that handling blooms roughly, restaging a bloom (other than replacing a bloom that was jostled out of place when moved), or otherwise mishandling an entry is not in the best interest of improving judging standards.

Being a member of a judging team is an honor as it indicates the show officials trust in your judgment. It should be a pleasurable experience for all members of the team. It is an opportunity

for each of us to increase our knowledge of the dahlia and to pass our hard-earned knowledge on to others. For most of us, dahlia culture is an avocation, a time to renew friendships, a cheerful time to create happy memories. Every judge can make this experience a pleasurable one or a time of frustration.

JUDGES' EDUCATION

Although it was not his intention when he wrote *The Judging of Dahlias in Shows*, Ward H. Cook, MD in fact introduced the first U.S. text for dahlia judging schools. The ADS adopted this initial text for dahlia judging and has revised it several times. The program currently has established four categories for judges:

- ★ Candidate
- ★ Accredited
- ★ Senior
- ★ Honorary Senior Judge

From the time of the first course, judges have progressed through these categories attaining each level by increasing their knowledge through refresher courses and judging in shows and trial gardens where they work directly with experienced judges. The senior judge, an experienced judge who assumes a leadership role, has the highest rank and best background in growing, showing, and judging.

The honorary senior judge category exists to acknowledge those judges who have served many years but, because of extenuating circumstance, are no longer active. On occasion, however, they may still judge or assist in some capacity, but cannot judge ADS bench evaluation entries or in the trial gardens.

ADS JUDGES' RECORDKEEPING

The ADS has implemented an organized procedure for administering judges' education, including online material. Established dahlia conferences and local dahlia societies (if they are not a member of a dahlia conference) maintain their own programs. Flexibility for these organizations exists so that they can determine the organizational structure they desire and how their general judging programs will operate. A conference or society will then report directly to the ADS Judges' Accreditation Committee Chair (accreditation chair).

Each dahlia society that is not a part of a conference may start a training program by appointing an evaluation committee consisting of a senior judge as chair and two or more additional senior judges. Once an evaluation committee is established, the accreditation chair, whose name appears in the listing of ADS officers and major committees in each ADS *Bulletin*, should be notified of the composition of the committee. For recordkeeping purposes, changes in the composition of the committee should be reported promptly to the accreditation chair.

The evaluation committee chair of each conference and each non-conference dahlia society accepts responsibility for maintaining accurate records of all ADS judges within their area. They report to the accreditation chair information such as name

Federation of Northwest Dahlia Growers
Requirements to become a
ACCREDITED JUDGE

The Accredited Judge must be able to judge competently in *all* sections of a dahlia show.

To be considered for advancement to Accredited Judge status, a person must maintain membership in at least one local dahlia society, the Federation of Northwest Dahlia Growers, and the American Dahlia Society, and meet all of the following requirements.

____ year	____ year	____ year		

☐ ☐ ☐ ☐ ☐ ☐ ☐ ☐ ☐ ☐	1. Exhibit dahlias at ten ADS-affiliated dahlia shows.
☐ ☐ ☐ ☐ ☐ ☐ ☐ ☐ ☐ ☐	2. Judge dahlias at ten ADS-affiliated dahlia shows.
☐ ☐ ☐	3. Judge a minimum of three seedlings per year at the trial garden or on the seedling bench under the supervision of a Senior Judge.
☐ ____ year	4. Attend the accredited class on dahlia judging and classification.
written ☐ practical ☐	5. Pass written and practical tests.
☐ ☐ ☐	6. Attend a minimum of one Federation workshop per year.
☐	7. Successfully complete a special project under the direction of the judges' evaluation chair.

Example of the recordkeeping form provided to Federation of Northwest Dahlia Growers judges to track progress toward judging requirements.

deletions because of death, disability, lack of interest; transfers to or from other societies; and changes in addresses and phone numbers. In addition, the chair must provide an annual report to the accreditation chair of each judge's status, making sure to include full and complete information.

The ADS will annually publish the names of all candidate, accredited, senior, and honorary senior judges. The conference or society evaluation chair for judges will submit updates to the accreditation chair. Full directory information will include first name and surname, address, zip code, telephone number, email address, home dahlia society, and level of judging accreditation.

The current requirements for each judging level can be found on the ADS website.

CHAPTER 2
DAHLIA CLASSIFICATION AND EVALUATION CHARACTERISTICS

The judging of dahlias requires the evaluation of individual qualities of the dahlia. The ADS has selected several characteristics for evaluating all dahlias. These include form, color, substance, stem, foliage, bloom position, uniformity (bench), and floriferousness (trial garden). The ADS scorecards have been developed for judging fully double dahlias and open-centered dahlias and approved by the ADS Executive Board. Each characteristic has specific value assigned to it that judges will use in their evaluation.

The ADS scorecard is a tool used in the evaluation of new cultivars and is discussed in chapter 4. Judges must assess the severity of the fault within the range of points specified. Experienced judges should have acquired a good perception of point values by growing and exhibiting show-quality blooms, by possessing a knowledge of many cultivars in commerce, by recognizing grooming techniques of exhibitors, and by applying knowledge gained from other judges. In this assessment, judges should examine entries and make their decisions from approximately three to five feet. Many characteristics are more apparent from a distance of at least three feet (arm's length), especially balance and proportion. Faults apparent from a distance should be punished more severely than those requiring a close-up look.

The ADS has established descriptive definitions for each recognized form of dahlia. Refer to the current CHD for the most up-to-date definitions. These definitions have become a standard for judges to follow when evaluating dahlias. This system becomes a means of description and organization among dahlia growers and provides a visual image for ADS-recognized dahlia forms. However, within these classes a wide range of appearances often occurs because nature has put the dahlia form on a continuum.

DAHLIA CLASSIFICATION AND THE CHD

Dahlias are classified by size, form, and color to describe them and to create a framework with which to organize shows. Official classification is handled by the ADS; individual cultivars and their classifications are listed in the CHD. Ideal dahlia form descriptions (with pictures) are described beginning on page 18 in this book. See appendix A for more information on dahlia classification.

Dahlias that do not consistently win blue ribbons on the show bench are dropped from the CHD but will continue to be listed in the *Composite Listing of Dahlias*, an ADS publication. The composite is currently published every five years and lists every dahlia variety that has appeared in a CHD since 1976. Copies may be purchased on the ADS website; all dahlia judges and exhibitors should own the latest edition. Content of the composite is also available on the ADS website.

There are several ways a dahlia can get listed in the CHD.

- ★ A cultivar that wins two blue ribbons in ADS-affiliated shows in a year will be initially listed – or relisted after being dropped from the CHD.

- ★ A cultivar must win three blue ribbons during the last two-year period to remain listed.
- ★ A cultivar with a score of 85 or greater from either an official trial garden or seedling bench evaluation will automatically be listed for at least two years.
- ★ A cultivar that receives a classification committee change recommendation will be included in the CHD the ensuing year even if it does not receive two blue ribbons.

When entering dahlias in shows, exhibitors are to enter them as classified in the latest CHD. Varieties not listed in the CHD might be found in the composite listing. If a variety does not appear in either publication, the exhibitor should get it classified by the classification committee at the show. The committee may be "any senior judge," a listing of several senior judges from whom to choose, or it may be a formal committee sitting at a table. The committee structure varies by show and will be described in the show schedule. If an entry is classified by a committee — and the entry tag so indicates — the entry must be judged, even if the judging team disagrees with the classification.

Many hybridizers choose to enter their named seedlings in open competition to see how they perform against current varieties. When doing so, they should (but don't have to) get the entry classified by the classification committee. A seedling entered with only a number need not be classified by a classification committee unless required by the show schedule. However, if the hybridizers don't get it classified — and the judging team disagrees with the hybridizer's classification — the team may set back and not judge the entry.

Similarly, if an exhibitor enters a named variety that is not listed in either the CHD or the composite listing — and chooses to not get it classified by the classification committee — the judging team may set back and not judge the entry if they disagree with the exhibitor's classification.

Hybridizers entering seedlings in shows may not want the cultivar to appear in the following year's CHD. For instance, they may not plan to release it immediately, and its listing in the CHD would be awkward. If the hybridizer does not want her or his seedling to appear in the following year's CHD, they should:

Enter the variety with only a number and not a name.

Or

Enter the variety with a two-digit number preceding the name, for example, "15-Bob's Big Beauty."
And
Notify the classification chair.

Exhibitors and judges should not rely on commercial grower classifications or classifications that come from dahlia growers in a foreign country since classification can vary from country to country.

Because some dahlias grow differently in various areas, the ADS allows local conferences to maintain lists of varieties that have a different classification locally than that listed in the CHD. The differences are typically in size, but different climates may also affect form or color. These "supplemental listings" may be located in the CHD or local show schedules and take precedence.

Occasionally exhibitors disagree with a dahlia's classification as listed in the CHD. There is a formal process to request reclassification. The request can be made by any ADS member and is outlined in the CHD under "Cultivar Classification and Change Request Rules and Procedures." Exhibitors should continue to enter their dahlias as listed in the current CHD until any changes are formalized and published. Appendix D has information on classification change requests.

Most varieties receive their classification when evaluated in trial garden or seedling bench evaluation competition. Judges assign classifications to the entries and those that score at least

85 are listed in the CHD for at least two years. If they win blue ribbons on the show bench they will continue to be listed; those that don't win blue ribbons will live on in the composite listing.

FORM CHARACTERISTICS FOR FULLY DOUBLE DAHLIAS

Symmetry

Viewed from the front of the bloom, all dahlias should present a symmetrical, circular outline. From the center of a bloom, the ray florets that are similar in distance from the center of the bloom should be of equal length. The bloom should also be well centered on the stem so that it appears to enter the bloom near its center rather than to one side. Often, the length of the lower ray florets, especially of larger dahlias, will be greater than those on the upper half of the bloom. This condition is often referred to as bearding and is a fault by degree. Since symmetry is such an important visual feature, judges must carefully assess its part in the overall effect of a dahlia form.

The center of the bloom must be round and tight with the developing ray florets filling the center completely. The size of the center should be in pleasing proportion to the bloom itself so that it appears to center rather than dominate the bloom. The ray florets of a maturing fully double bloom on occasion may not completely meet at the center of the bloom resulting in a small hole; this condition is a minor fault and should not be considered the same as a blown center or a green center.

The blown center should not be confused with open-centered dahlias. A bloom with a blown center – where pollen-bearing disc flowers are partially or entirely evident – should be eliminated from further competition. Possible faults of centers also include centers without floret color (hard and green), oblong, depressed, protruding, or presenting a flat face.

Contour

Judges must examine a fully double bloom by looking at the overall shape and build. Ideally, the bloom will appear round from the face. The form will generally be spherical with the receptacle functioning as the bloom's center, except for waterlily, incurved cactus, and stellar formations. Judges must also examine more than the outline of the bloom; they must inspect the arrangement, placement, and uniformity of ray florets.

The ideal bloom will have a small, closed, well-proportioned center with developing ray florets that appear to spiral in a regular formation. These ray florets will uniformly and gradually increase in size from the center to the most mature ray florets at the bloom's back. Faults occur when there are gaps, an irregular placement of the ray florets, or irregular formation of the ray florets. A faulty contour on an overly mature bloom usually causes an asymmetrical appearance and can affect bloom symmetry. The faulty build of a bloom will appear shaggy and/or irregular in formation; it may affect uniformity.

Development

For the center of the bloom to appear pleasing, it must be in proper proportion. A bloom that is not fully mature will have a larger center and be out of proportion; a center in proportion will provide a focal point, but at the same time, blend in with the surrounding ray florets and general bloom presentation. Usually, an immature (green) bloom will lack depth but will also display excellent color. The judge must remember that the mature bloom at the peak of development with minor faults merits first choice over an immature, "clean" bloom.

Trueness to Form: The ADS has established descriptive definitions for each recognized form of dahlia. These definitions have become a standard for judges to follow when classifying dahlias. This system becomes a means of description and organization among dahlia growers and provides a visual image for ADS-recognized dahlia forms.

For classification purposes, the descriptions that follow represent the ideal for a dahlia form. Few cultivars will have ideal form because of the many variations of form in the dahlia domain.

➲ *See appendix E for a complete listing of dahlia sizes.*

FORM DESCRIPTIONS OF FULLY DOUBLE DAHLIAS

Formal Decorative

Ray florets are generally flat, broad, and smooth in a regular arrangement that gradually recurve toward the stem. Each ray floret row should convey uniform and regular size arrangement with the tips preferably round or extending to a slight point. The ideal depth is three-quarters the diameter of the bloom and should not be greater than its diameter.

Formal Decorative

Informal Decorative

Ray florets may be twisted, curled, or wavy, and of uniform size in regular or irregular arrangement. They may be partially involute or revolute, but no portion should be fully revolute or involute except at the tip of the ray floret. The ideal depth is three-quarters the diameter of the bloom and should not be greater than its diameter.

Informal Decorative

Semi-Cactus

Ray florets are broad at the base, straight or incurved, and reflex toward the stem in a regular and uniform arrangement. Ray florets will be revolute for approximately half their length and fully revolute (touching or overlapping) for at least one-fourth their length. The ideal depth is three-quarters the diameter of the bloom and should not be greater than the diameter.

Semi-Cactus

Cactus

Cactus

Ray florets are narrow at the base, straight, uniform in length and reflex toward the stem, radiating uniformly in all directions from the center. They will be revolute for the majority of their length and fully revolute (touching or overlapping) for approximately half their length. A depth of about three-quarters the diameter is ideal for this form.

Incurved Cactus

Incurved Cactus

Ray florets are pointed, narrow at the base, uniform in length and arrangement, revolute for the majority of their length and fully revolute for at least half their length. They will curve upward to the bloom face, but may also arc downward and then up, or they may swirl horizontally around the bloom circumference, usually exposing the involucre in either case. Depth must be at least half the diameter.

Laciniated

Laciniated

Ray florets are split at the tip with the number and depth of the splits varying according to the size of the ray floret. The portion of the ray floret with the splits should uniformly twist or curl whether involute or revolute. The less mature ray florets should possess splits but do not have to twist or curl. In the ideal laciniated dahlia, the majority of ray florets will be uniformly arranged with a split, twist, or curl that conveys an overall fringed or frilled effect. Ray florets, slightly involute or revolute, that split regardless of length and remain generally flat are a major fault. Ray florets with a shallow notch or cleft and possessing little twisting to the extreme portion (tips) should not be classified as laciniated. The ideal depth is three-quarters the diameter of the bloom and should never be greater than its diameter.

Ball, Miniature Ball, and Pompon (top) exhibit fully involute ray florets (bottom).

Ball

Blooms are ball shaped with uniform florets, involute for most of their length, and fully involute for more than half the length; the ray florets reflex completely toward the stem. Ray floret tips in cross-section should be circular and exhibit little distortion, completely filling the floral head and are either round or blunt without notch or point.

Miniature Ball

Except for size, miniature ball and ball dahlias possess the same form features.

Pompon

Ray florets are involute for the whole of their length and fully involute for over half their length. Except for size, they are similar to the ball description.

Stellar

Ray florets are long and narrow along their length with pointed tips preferred; like a star, they exhibit space

Stellar

between the ray florets in each of the rows and possess a uniform, regular arrangement. All the ray florets will have a partially involute form in an approximate "U" shape in cross section for the majority of their length and will recurve toward the stem; the bloom depth should be greater than one-half the diameter but should not be greater than its diameter.

Waterlily

The bloom should be fully double and symmetrical; the side view should be flat to saucer-shaped in arrangement, and the layer of florets should be open faced to give a delicate appearance. The center should be closed and dome-shaped, breaking gradually to four to seven rows of fully developed outer florets. Outer florets should be broad and slightly cupped with rounded tips. A bloom position at 45 degrees is preferred, but a top-facing bloom is acceptable. In petite waterlily cultivars (from 1 to 3 inches in diameter), the depth should contain fewer rows of ray florets in order to represent the waterlily form accurately. The depth should never be more than one-half the diameter of the bloom.

Fully Double Novelty

These dahlias will display distinct and different form from present classifications. The new formation must display good symmetry, including a closed, tight center, ray florets uniformly shaped and/or arranged differently from other fully double form characteristics.

Waterlily

Fully Double Novelty

Fully Double Form Summary

Ray florets should be of consistent size and form as they spiral from the center of the bloom. Throughout the bloom, a uniform spacing of ray florets without apparent gaps should exist. The fully mature bloom will convey symmetrical form relative to the center of the bloom from its face, from side to side, and from top to bottom. Each side of the bloom will also have a symmetrical appearance. The bloom will have a sufficient ray floret count for its form. The center will be round, tight, and in proper proportion to the size of the bloom.

Fully double form faults

- ★ Center is not round, is too large, too high, depressed, open, loose, elongated, oblong, misshapen, green, weak, or damaged.
- ★ The symmetry of the bloom is affected if it is elongated, lopsided, or bearded when viewed from the front or side.
- ★ The front or a portion of the bloom tends to be flat and does not exhibit uniform symmetry.
- ★ Inconsistent ray floret form (mixed form).
- ★ Inconsistent arrangement of ray florets (gaps).
- ★ Apparent loss of or dropping ray florets.
- ★ Sparse ray floret count.
- ★ Ray florets not in proportion to size of bloom. For example, large, BB-sized ray florets on a miniature bloom which leads to "coarseness."
- ★ Visible petaloids or bracts.
- ★ A minor fault exists when a cultivar's center does not fully close but reveals a small hole. This condition is not an open or blown center.

FORM DESCRIPTION FOR OPEN-CENTERED DAHLIAS

eony

Peony

At least two, but not more than five rows of ray florets surround the disc. Ray florets are broad, generally flat to slightly cupped and evenly spaced in a flat plane. Ray florets are equal or nearly equal in length from the bloom face to the back, not recurving or reflexing to the stem. Ray florets adjacent to the disc may be twisted or curled, but if present should be symmetrically arranged around the disc. Ideal depth is less than one-third the diameter. Disc flowers with 0 to 3 rows of pollen are desirable.

nemone

Anemone

A dome of elongated tubular disc florets are surrounded by one or more rows of ray florets. The ray florets should be uniform in shape, size, and formation and be regularly arranged around the disc florets. The ray florets should be visible from the face of the bloom, creating a pleasant and balanced framing of the dome. The disc florets should also be uniform in shape, size, and formation and be fully developed. Each disc floret should be entirely tubular or predominately tubular with a fringed tip.

lovelty Open

Novelty Open

Dahlias with characteristics distinct and different from the present classifications are classified as novelty dahlias. The center should be open with the disc center in proper proportion to the ray florets.

Collarette

A bloom will have a single row of flat or slightly cupped ray florets arranged in a flat plane; the ray florets will uniformly overlap, preferably in the same direction with no gaps. The petaloids surrounding the disc should be approximately one-half to two-thirds the length of the ray florets. Eight ray florets are most desirable. A round floret tip is preferred. Disc flowers with 0 to 3 rows of pollen are desirable.

Collarette

Orchid

A single row of uniform, evenly spaced ray florets arranged in a flat plane surround the disc flowers. The ray florets should be straight and involute for two-thirds or more of their length and fully involute for at least one-third their length. Eight ray florets are most desirable. Disc flowers with 0 to 3 rows of pollen are desirable.

Orchid

Orchette

A bloom combines the involute characteristics of the orchid dahlia form and the petaloid characteristics of the collarette form. Ideally, eight ray florets uniformly surround the disc flower in a flat, even plane. Ray florets should be straight and involute for at least half their length and fully involute for at least one-fourth of their length. Ray florets should possess a corresponding petaloid at their base. Petaloids should be of a length which gives visual impact and not so small that they are not easily discernible at arm's length. Petaloids may lie flat within the base of

Orchette

the ray floret or slightly rise and/or cup upward toward the center of the bloom. Disc flowers with 0 to 3 rows of pollen are desirable.

Single

Single

The bloom has a single row of flat or slightly cupped ray florets arranged in a flat plane. They uniformly overlap, preferably in the same direction with no gaps. Eight ray florets are most desirable. Disc flowers with 0 to 3 rows of pollen are desirable. A round floret tip is preferred. The bloom is over two inches in diameter.

Mignon Single

Mignon Single

Mignon single dahlias possess the same formation as single dahlias, except they have round ray floret tips. Disc flowers with 0 to 3 rows of pollen are desirable. The bloom is up to two inches in diameter.

Open-Centered Form Summary

Ray florets must convey uniform size, form, and arrangement in a flat plane. Ideally, mature cultivars with an open center should exhibit eight ray florets with 0 to 3 rows of pollen showing.

Open-centered form faults

- ★ Ray florets of inconsistent size, form, or arrangement.
- ★ Bloom not in a flat plane.
- ★ Immature ray florets.
- ★ Browning pollen.

COLOR

The ADS has selected color as one of the two most heavily weighted characteristics for evaluating dahlias. The ADS recognizes 15 color classifications of dahlias including white, yellow, orange, pink, dark pink, red, dark red, lavender, purple/black, light blend, bronze, flame blend, dark blend, variegated, and bicolor. Each color class is defined in the current CHD published by the ADS. Numbers and/or color names identify where judges should place a cultivar. The color chart chosen by the ADS will serve as the official authority for cultivar classification.

The competent dahlia judge should approach each of the 15 color classes without prejudice, recognizing that no color class has any greater merit than the others. A wide range of color variety will exist within these arbitrary color classifications. In any ADS-sanctioned show, exhibitors will enter their dahlias according to the most current CHD.

Color Terminology

The language of color includes hue, value, and chroma. Every color sensation unites these three separate qualities.

Hue simply names a color. It allows one to distinguish one color family from another, such as red from yellow. Primary spectrum colors include red, yellow, and blue. According to the pigment theory, they cannot be broken down into other hues. Secondary spectrum colors mix two primary colors together, and tertiary spectrum colors mix a secondary and a primary color. A hue consists of the above combinations of pure color without any value admixture.

Value indicates the lightness and darkness that any color may possess, depending on the amount of white or black that is present in the color of a dahlia. The value will range from spectrum colors mixed with white referred to as tint, also known as high or light values, to shades mixed with black, also known as dark values. The combination of tints and shades will produce grayed spectrum colors or tones. Judges should refer to value of colors as lighter, darker, or grayed rather than tints, shades, and tones.

Chroma is the strength or weakness of a chromatic color. Its quality of color allows one to distinguish a weak color from a strong one. One may also determine by degree how far a color sensation departs from white or gray. In terms of chroma, color is often described as weak, moderate, or strong. Intensity of color is an inherent quality that distinguishes between a weak and strong color. All spectrum hues possess full intensity although some appear weaker than others. For example, a strong, brilliant red possesses a saturation or intensity that is much greater than a weak, dull red. In this instance, the chroma passes from the strongest red in the chroma scale through medium red to a weak, grayish red.

In judging dahlias, a lightness of a color is not necessarily a negative characteristic. For instance, if color is evenly distributed in both a yellow pastel dahlia and a more saturated yellow one, the quality of color is not better in one than in the other. If an exhibit shows a purity of color and possesses an evenly distributed color, it deserves a premium rating. Although lighter tints of any color will have a reduction of color saturation caused by the amount of the admixture of white, the color remains pure and decreases the intensity of the chroma. Conversely, if the color value is altered by an admixture of gray, the color loses both color intensity and color purity; this is a negative characteristic.

The neutral colors of black and white do not appear as a part of the spectrum colors. Yet today, for descriptive purposes both are accepted as colors. A true white has no color factors. Black, contrary to tone, is pure in color and possesses depth and richness of color. It is the admixture of shade and tint that produces dullness (tone).

To summarize, pure color combines to four secondary forms: white and black create gray; a hue and white create tint; a hue and black create a shade; and a hue, white, and black create tone.

Judging Color Quality

Most dahlias have some gray in them. Many of these cultivars appear quite outstanding in the garden, especially when the sun shines directly on them. Frequently, however, the graying factor contributes to a dirty color quality. When a slightly grayed dahlia is contrasted with a whiter dahlia, this becomes much more evident, especially indoors where the surface sheen is less of a factor.

Imperfections of Color

Conspicuous green bracts at the base of a ray floret also affect a bloom's visual appearance, as do individual ray florets of a color different from other florets. Judges must score these and other genetic characteristics down. In both the trial garden and on the bench, wolf petals are severe faults. On the show table, a different colored streak in a floret will distract from the inherent beauty of the bloom while in the trial garden or on the seedling bench, judges should penalize this flaw more severely. In either case, a judge must assess the degree of fault. In self-colored dahlias, the spread of the color on the ray florets should be evenly distributed without a blotchy, streaked, veined, or granular appearance at approximately arm's length.

Blushes in Dahlias

Judges often have a difficult time determining when a cultivar exhibits a blush and when it is a blend. In each instance, more than one color appears. To add to this problem, climactic conditions and the time of season can also play a role in the cultivar's appearance. Slight blushes can occur in all colors and often appear differently depending on where they are grown or can disappear completely late in the season.

A light, uniformly tinged coloration on florets represents a blush. It is imperative, however, that a judge be flexible in blush and blend evaluation since many different conditions exist. Judges must recognize that the presence of a uniform blush is not a color fault. Judges should classify cultivars with a blush in the solid color classes. Leniency should be the rule when judges evaluate cultivars that display a blush since the blush in a cultivar can vary substantially.

Special conditions also exist for the white dahlia. Judges need to remember that a slight blush of ivory, cream, lavender, or pink often enhances a white bloom. These blushes do not decrease the purity of the bloom's appearance.

Light and Dark Blends

In blends, two or more merging colors on either the face or reverse of the ray floret OR two or more contrasting colors on the face or reverse of the ray floret should be apparent. If a variety has tints, tones, and/or shades of a hue that vary by degree in the same color class, judges will classify it as if it were one color. For instance, if straw yellow, gold, and pastel yellow all combine in a variety, since all belong to the yellow color class, judges will classify this variety as yellow and not a blend.

A blend must contain more than one color with portions of these colors appearing separate and distinguishable.

Top: Light Blend
Bottom: Dark Blend

Light blend varieties consist of lighter tints and tones of pink, yellow, lavender, and other pastels; in addition, there are two-toned varieties of pastel tints and tones that have central ray florets which are markedly different in color from the marginal ray florets.

Dark blend varieties embody colors of lower brilliance that may either merge gradually or contrast; these colors appear in the latest CHD and among other colors will include any of the dark reds and purples. At times, a dark blend variety will appear quite brilliant, especially when it has white as a contrasting color. Like the light blend, the dark blend may also be two-toned if the variety possesses a darker color.

Since white is a component of many blends, both light and dark, white is not a factor in determining whether a variety is a light or dark blend.

Blend color classes allow for a separation of colors and a means of describing multicolored varieties. For the most part, a judge must assess the lightness or darkness of color to determine whether a variety is a light or dark blend. As a guideline for assessing a blend, judges should distinguish whether a variety exhibits the qualities of a blend from arm's length. From this distance, if it is a blend, it will possess two or more distinct colors that are not a part of the same color class.

Colors of one variety may contain both light and dark blend colors. The judge will then determine the blend classification by establishing its dominant color and assigning its color class accordingly. If the dominant color is a dark blend color, the cultivar will be a dark blend; if the dominant color is a light blend color, the cultivar will be a light blend.

Flame

Flame dahlias have a separate class from other combinations. The striking contrast of fiery red, yellow, bronze, and orange attracts both the public and exhibitors alike. The flame dahlia will simulate flames of fire by the merging of colors or by displaying a definite color break at greater than one quarter the length of the ray floret. Judges should place a variety with bicolor characteristics in the bicolor color class; otherwise, they should classify the variety as a flame. Like the flame metaphor, the actual color distribution on a floret in a variety may reveal a significant difference in color arrangement from another variety. Flame dahlias must contain a red, bronze, or orange hue that will combine with yellow. Check the CHD for current flame color chip numbers.

Variegated Dahlias

Variegated dahlias contain two or more distinct colors on the face or reverse of a ray floret including stripes, splashes, dots, flecks, and narrow lines. Variegation is inherently random and is not likely to be totally uniform. For orchid and orchette cultivars, judges will determine the variegated quality by viewing the back of the floret. A variegated cultivar that reveals sharply contrasting colors in adequate proportion should rate higher in color than one with poor contrast. Likewise, one with a more even distribution of variegation should rate higher than one with large blotchy spots or uneven distribution. Unevenly distributed flecks throughout the florets are a minor fault while a solid-colored floret on a variegated cultivar is a major fault.

Bicolored Dahlias

Bicolored dahlias contain solid-colored ray florets with tips of a different color or stripes of a different color extending along the ray floret length. The ideal bicolor will have consistent, uniform tips or stripes of distinct, contrasting color with little bleeding. If a variety does not possess this distinctness, judges should classify it as a blend.

In a bicolor, all the ray florets should be uniformly marked with a delineated proportion of both colors. A floret of a bloom may be tipped white or may have a longitudinal stripe. Bicolors should be one-quarter to one-sixth tipped. A bloom with only a slight contrasting tip is considered a self-colored (single colored) bloom.

The striped bicolor, on the other hand, which extends the length of the ray floret should contrast from one color to the other by one-quarter to one-half the width of the ray floret. Cultivars classified as bicolors occasionally produce self-colored blooms or blooms showing only a few bicolored ray florets. Judges should eliminate self-colored blooms and severely fault an entry if bicolored and self-colored ray florets intermingle.

For classification purposes, bicolored dahlias should have a sharp, clean-cut separation of contrasting color with minimum bleeding of one color into another. All cultivar tips or stripes, white or otherwise, must contain clear bicolor characteristics.

Color Insights

Soil ingredients, shade conditions, fertilizer, and climate may affect color quality. In the extreme, a variety grown under different conditions may display such vastly different colors that judges will question whether it is the same variety. A bloom that burns or fades, whether it is the entire bloom or merely the tip, should be faulted accordingly. When judging the color classes in competition, judges should also observe and fault other conditions such as water spotting, spray deposits, mildew, and damage caused by travel or rough handling. Judges should penalize pest injury and discoloration to ray florets under either form or color but not both characteristics. Browning or fading directly affects color; a judge seeing a hole in a floret would penalize form.

The quality of color may be notably enhanced by the texture of the ray florets. This term describes the surface of the floret where it may have a cloth-like, dull texture or a velvety, silky, satiny, or sparkling texture that may enhance the bloom's appearance.

Color Bias

If a class permits cultivars of more than one color to be shown in competition, judges should be especially careful to avoid discrimination against colors which to them may be personally less pleasing, but which may properly rate high in the cultivar's own color class. When blooms of more than one color are exhibited together in any multiple-bloom class; color harmony or pleasing contrast is always preferred.

Color Determination

Judges who determine color should use the color guide adopted by the ADS, examining and matching the entries in natural light. Determine color in shade, but never in direct sunlight or artificial light.

When identifying color for descriptive purposes, judges should consider at least three blooms of the same variety to accurately determine the color(s) of a cultivar. They should also determine color by evaluating blooms that are at the peak of development.

When a variety contains several colors, judges should determine the dominant color first and then the secondary ones. The specimen should be held against the color chip or vice versa. Any adjacent color can affect the apparent color of both the dahlia and the standard. A judge may pluck a floret at the equator of the bloom, which represents the majority of ray florets, and use it to identify the color precisely.

Colors are occasionally difficult to match. One of the reasons for this problem is that often a subtlety of color exists. At first, the eye may see only one color when in fact two closely related colors exist. Be aware that some judges are inherently better at discerning color.

FANCY PANTS

Classification procedures for color is technical. It is fully documented in CHD. In addition, there are tutorials on the ADS website. Briefly, this is the process that leads to the precise descriptions of a dahlia's color in the CHD. For instance, one of the more complicated color descriptions is for Fancy Pants: LV5/lb/yl7/lv5-lb/yl7/lv5. The candidate judge is not expected to master this process; accredited and senior judges typically use the process only when judging in trial gardens or in ADS bench evaluation situations.

Color Summary

A bloom must have uniform color distribution. Its individual ray florets will have a clean, appealing surface texture with a lustrous appearance. All cultivars described as blends will exhibit a uniform color distribution either by merging gradually or by a pleasing contrast.

The ideal bicolor will have consistent, uniform tips or stripes of distinct, contrasting color with little bleeding. A floret of a bloom may be tipped white or another color or may have a longitudinal stripe. Bicolors should be one-quarter to one-sixth tipped. A bloom with a slight contrasting tip is considered a self-colored bloom. Judges should classify cultivars that possess two or more contrasting, separated colors as a bicolor if the ray florets display bicolor characteristics.

The ideal variegated cultivar has a uniform base color(s) throughout with narrow stripes, splashes, lines, flecks, or dots of contrasting color(s) uniformly distributed on all ray florets.

Color faults

★ Major faults include cultivars that are dull, splotchy, streaked, have distracting bracts, an occasional ray floret of dissimilar color, and poor distribution.

★ A bicolor or variegated cultivar with some or all ray florets of solid color, with uneven color distribution, or without adequate color contrast.

★ A bicolor with poor separation of color, excessive bleeding, and uneven color distribution.

★ Color blemishes including fading, burning, spotting, bruising, insect damage, mildew, and spray deposits.

SUBSTANCE

Ray florets and foliage should appear fresh, clean, crisp, springy, and firm. They are free of spotting, wilting, mildew, burning, dropping ray florets, insect injury, or mechanical damage. A light touch with the back of your hand is permissible to help evaluate the substance of a bloom.

Substance faults

★ Ray florets are soft, spotted, wilted.

★ Ray florets do not hold form.

★ Blooms have a loss of or dropping ray florets.

There are several factors that can affect the substance of a dahlia including:

- ★ Lack of adequate watering.
- ★ Picking dahlias during the heat of the day.
- ★ Insect damage.
- ★ Failure to adequately condition blooms after picking.
- ★ Hot weather.
- ★ Genetics.

Example of soft, browning ray florets indicative of poor substance.

Note that the addition of a floral preservative to an exhibit's water may enhance an entry's substance. This is an acceptable conditioning practice.

STEM

An ideal stem is straight, round, smooth, and of adequate diameter and length from the bloom to the first pair of true leaves; the stalk continues straight to the top of the exhibition container. The length of stem must be in proportion to the diameter of the bloom and the size of the foliage. A small diameter stem on a large bloom does not have good proportion nor does a small bloom with a large diameter stem. There should be an attractive relationship of the three components: bloom size, stem length and diameter, and size of foliage.

Example of stem out of proportion to the diameter of the bloom.

Stem faults

- ★ Cultivars with stems too long (willowy) or too short (crotched).
- ★ Stems not straight.
- ★ Stems ridged.
- ★ Stem diameter too large or small.
- ★ Stems weak for size of bloom.
- ★ Stems not smooth or round.
- ★ Stems are rubbery.

FOLIAGE

Leaves should present a pleasing, uniform color; each pair of leaves should be equal and opposite on the stem or stalk and ideally free from damage. The size and shape of leaves, either simple or compound, should be uniform in appearance. The total spread of the foliage should be greater than the diameter of the bloom; a longer stem should have foliage with somewhat greater spread. Regardless of bloom size, foliage can be either too large, presenting a squat appearance to the overall proportion, or too small, giving the appearance of a longer, awkward stem. There should be minimal scars from disbudding and disbranching. The foliage of a cultivar should be in proper proportion and frame the bloom as it faces a person.

Example of foliage which is too large relative to the size of the bloom.

Foliage faults

- ★ Leaves appear diseased.
- ★ Leaves stair stepped.
- ★ Leaves are too small or too large relative to the size of the bloom and length of stem.

★ Leaves are not generally flat or of similar size or form.

★ Leaf color is not uniform.

★ Foliage shows obvious damage.

★ Foliage shows spray deposits or mildew.

★ Leaves are pale, blotchy, coarse, soft, crinkled, folded, or yellow.

★ Foliage aligned to the front and back of bloom does not properly frame it.

★ Late disbudding and disbranching hinder the visual appearance of foliage.

BLOOM POSITION

The preferred bloom position is approximately forty-five degrees relative to the stem. However, for pompon, miniature ball, ball, and waterlily forms, a top-facing bloom is acceptable. Many larger blooms that are top facing, while not displaying the preferred position, may greatly improve the form of the bloom by reducing apparent bearding.

Bloom position faults

★ Down facing is a major fault with all types of dahlias.

★ Side-facing or top-facing is a fault except as noted above.

Example of a side-facing bloom.

UNIFORMITY

All blooms of a multiple-bloom entry must be mature and exhibit uniformity for all evaluation characteristics. Judges must evaluate the uniformity of each characteristic. Judges will make deductions for each characteristic lacking uniformity.

Uniformity faults

- ★ Lack of uniform color and form are faults by degree.
- ★ Variance in bloom sizes.
- ★ Variance in bloom position.
- ★ Varying stem lengths.
- ★ Leaf shapes not uniform.

FLORIFEROUSNESS IN TRIAL GARDEN EVALUATION

This characteristic is not considered when judging in shows or in ADS seedling bench evaluation; it is only evaluated by accredited and senior judges in trial garden judging (see chapter 4). There are adequate blooms, branches, and buds based upon the size of the bloom. The plants are vigorous; mature blooms are similar and uniform for all evaluation characteristics.

Floriferousness faults

- ★ Overall, too few blooms.
- ★ Blooms too late.
- ★ Too few quality blooms.

Dahlia foliage comes in many different shapes, sizes, and colors.

CHAPTER 3
JUDGING DAHLIAS AT SHOWS

The ADS expects credentialed judges to judge ADS participating or affiliated society shows.

Fairs, Flower, and Horticulture Shows: Dahlia judges are often asked to judge at fairs and horticultural shows. These events usually have fewer classes of dahlias than regular society shows and may contain other horticulture entries. Since ADS judges are often asked to assist in writing schedules, they should use proper dahlia terminology to help educate the public. In small shows, judges may also be asked to judge other flowers and even other sections. They should judge dahlias according to the latest information available and research proper judging procedures for other floral or plant sections.

Ethics of Judging: The practice of judging is a teaching opportunity for the next generation of judges. Polite, instructive critique should prevail. Rules of good conduct define the ethics of judging. Ethical conduct is based upon honesty, empathy, integrity, sincerity, courtesy, intelligence, and wisdom. Judges have a particular duty to follow the principles of ethical behavior as outlined in chapter 1.

GENERAL RESPONSIBILITIES BEFORE A SHOW

Judges should:

- ★ Graciously accept judging assignments.
- ★ Answer judging invitations promptly and honor all judging commitments or notify show officials promptly if unable to attend.
- ★ Upon arrival at the show, immediately inform the judges chair that you are ready to judge. Determine which sections you will be judging and with whom (understanding that it may change by the judges' meeting).
- ★ Familiarize yourself with the physical show layout. Look over each section so that you will have a working knowledge of the caliber of the show and the general condition of entries. This preliminary activity will become the basis for why an exhibit should receive a blue, red, or no ribbon at all. The caliber of entries depends upon external factors, particularly weather.
- ★ Review the show schedule and rules so that you are familiar with the show rules. Remember that show committees vary rules according to their society preferences and area conditions.
- ★ Find the section(s) you will be judging and start to organize the flowers in form and color classes so they are ready to be judged.
- ★ Attend the pre-show judges' meeting for team assignments and any special instructions for the show.

- ★ Stay with your team until the judging assignment is complete.
- ★ Ensure that you are familiar with the show's clerking system.

ESTABLISHING A COMFORTABLE TEAM

After the judging chairperson gives the team their assignment, team members should greet and welcome each other by introducing themselves and stating where they are from and the society in which they are a member. Be alert as to how each person introduces themselves so that you are aware of the first name they wish to use. Doing this introduction will put the entire team more at ease.

The most important person on the judging team is the clerk; many times this show will be the first time that person has clerked. Clerks represent a lifeline for dahlia societies because they are the future judges. Ensure that the clerk understands the local procedures, how to mark the clerk sheets, and they are comfortable with their task. Everyone on the team will have to be sure the clerk fills out the clerk sheets correctly; everyone signs the clerk sheets when the judging is complete. Be sure to give the clerk attention during all phases of the judging process to ensure accuracy. By providing basic protocols, all team members should feel comfortable with their assignments.

Team leaders can designate the judging responsibilities for each team member based on their individual experience level. For example, who will place ribbons (judges, not clerks, do this) and who will check the CHD for proper placement of all the dahlias in the section you are about to judge. Remember this is a learning experience for candidate and accredited judges to learn how to use their tools and reference materials.

One vocal judge should not take over the judging. Ensure that each member of the team gives their recommendations about each entry so a consensus is formed. Do not override your team's decision just because you don't agree; it doesn't mean they are incorrect.

HOW TO DEAL WITH JUDGING YOUR OWN FLOWERS IN A SHOW

Team members who have an entry in any of the sections to be judged need not announce, "I have an entry in this section so I can't judge!" They should simply keep quiet when discussing a certain entry. If asked directly for an opinion, they should quietly say, "Sorry, I have an interest in one of these entries." Your team will understand and will come to a decision without your input and will then move on to the next color class. Stating only "an interest" in a particular cultivar could mean it's your favorite flower, it is your friend's origination, or for any combination of reasons you cannot be impartial in your judgment of that flower (not just that it is your flower, which puts a lot of stress on the rest of the judging team). If one or more of your "interests" moves up in the judging process, see if you can swap judging assignments with another judge, so that every entry gets a full, impartial assessment. Once your entry loses, it's back to work with your team. This is true even if you are the senior judge and the team leader.

TEAM LEADER RESPONSIBILITIES

Senior judges should provide mentorship, enabling the growth of future judges.

It is expected that team leaders establish an atmosphere of participation, cooperation, and input in the decision-making process; it is a team effort. When it's all over, everyone on the team should be able to explain the judging decisions the team has made.

It is a good practice as a team to reread the definitions of the forms that you will be judging before you start. The finer classification points will then be fresh in mind for all involved.

The team leader should set the example of courtesy, a stable temperament, and fairness, making sure that each member participates in the decision-making process. Ensure that each judge has ample opportunity to view each exhibit and do not monopolize the best view.

The team leader should actively participate in discussions but not dominate. They should be positive in their assessments and encourage everyone to give their rationale for their choice, letting them know at the start that it is okay to make comments and ask questions. They should encourage team members to request clarification if at any time they do not understand what is being discussed. Encourage clerks to participate; endeavor to point out good points and faults. Any decision must be a team decision.

JUDGING PRACTICES

Always judge every entry in a class. If a section looks incomplete, look around to ensure some entries are not on another table around the corner. Look carefully at each entry and then begin the elimination process. Make sure to assess the positive qualities as well as the negative. Judges must be alert to the passing of time and act in an appropriate fashion. It is, however, better to move slowly and thoroughly than to judge too quickly. Exhibitors deserve that their entries be given the attention they deserve.

Judging is a learning experience; it is up to each judge to educate through practical experience. For example, explain why one fault is more serious than another, why a fault is more severe on the face of the bloom rather than on the back, and then continue through the other characteristics. Show your knowledge, but with humility. Do not take on the air of knowing more but do reveal your experience and competence. It takes good demeanor to put judging into perspective so that the process becomes an enjoyable experience rather than a stressful one.

Misplaced entries may be moved to your section after you have judged; check with your local show chairperson on how to proceed.

Judges must be fast, efficient, and remain on task until the judging assignment is complete. Do not wander away from your team!

➲ *See Appendix N for a sample judging procedure.*

AFTER COMPLETING THE ASSIGNMENT

Upon completion of an assignment immediately review and sign the clerk sheets, ensuring they have been completed accurately. It is the judges' obligation and responsibility (not the clerk's) to ensure they are accurate.

Judges must make sure they attend to the show chair's requests such as carrying the winning entries to the head table and returning remaining entries to their proper place; leave a good display for the public after judging is completed.

Before the team disbands, check with the judges' chair to see if the team is needed for another judging assignment.

Candidate and accredited judges need and want honest input. When senior judges instruct in a friendly and positive manner, they will want to know how you view their judging. They are on a long journey and know they have much to learn before they reach the end. By being friendly and courteous, you model how all judges should behave when they are in your position. This extra effort on the part of the senior judge will produce more effective and positive judges in the future.

RESPONSIBILITIES OF JUDGES TO THE PUBLIC AND OTHER SHOW EXHIBITORS

Be alert to the public's interest and answer questions in a friendly manner. Don't be too harsh if they get in the way or touch the flowers, rather encourage their interest; they could be your next new members and hooked for life.

Don't make adverse remarks about the show, the flowers, other judging teams, or their decisions. Avoid standing around the head table and criticizing or second guessing the decisions of the judges.

Be able to explain your team's decision to any exhibitor if asked. Be most careful in the handling of the entries.

Don't judge too harshly – whether judging on the horticultural bench or the seedling bench; encourage exhibitors and hybridizers. We don't need to be punitive, just fair. Written comments on the entry tags can be helpful.

JUDGING TECHNIQUES

There are two ways that judges can evaluate blooms. The first is to compare each bloom against perfection. In this method, they compare each bloom against a perfect characteristic or quality. They then determine how far a characteristic is from perfection and convert the degree of fault to a number. For each characteristic on a scorecard, they examine objectively and deduct from the designated number of points. This technique takes a great deal of experience and mandates that judges grow, study, and practice judging dahlias at every opportunity. Experienced judges will employ this method when they judge entries on the ADS Seedling Bench or in trial gardens as discussed in chapter four.

The second way to judge, and the focus of this chapter, is to compare the characteristics of one bloom against another. Judges use this method when judging in shows, where they determine which entry is the best on the show table. Exhibitors refer to a show schedule and exhibit their entries in appropriate classes. When a class has more than one entry, judges carefully examine all the entries and then prioritize them, determining the ranking of all entries by evaluating the attributes and faults. The process continues until the best entry in the class is chosen. Judges have done the ranking by comparison. In this method, judges should be expedient, while making sure to judge every entry in the class.

Prior to Judging

After arriving at a show, judges should evaluate the quality of each division and the entire show to determine what the overall condition of the entries in the show. External conditions, such as intense heat, wind, or recent rain, can alter show quality substantially. Typically, three judges (one a team leader) and a clerk comprise a judging team. Before judging, the team should make sure the entries are arranged in proper order.

During Judging

As they judge, team members should check each entry for correct name, spelling, and classification and make corrections on entry tags if necessary. They should then carefully examine each entry and select the one most worthy. The clerk will record the judges' choices on the clerk sheets. Judges should hang ribbons and mark entry tags as appropriate; clerks should not be expected to perform those duties.

SHOW COMPETITION JUDGING FUNDAMENTALS

Judges should commit to doing a thorough, orderly, and methodical evaluation of all entries. It is ideal to initially examine entries from the furthest possible distance (at least arm's length). Balance, proportion, and major faults are thus easily and efficiently assessed. The severity of a fault is proportional to the distance from which it is apparent. A fault apparent from six feet is more severe than one apparent from six inches. Faults on the face of an exhibit should be penalized more severely than those on the sides and back.

Judges should examine all entries from all angles. Taller entries may need to be placed on the floor to judge the tops and

centers. Both sides should be evaluated as well as the back and underside of each bloom.

It is advantageous for judging teams to go through the process together while discussing observations. Judges should feel free to express and discuss their opinions with one another. This thorough discussion and its conclusions will enable any judge on the team to rationalize decisions that were made collectively if asked or challenged concerning the placement of an exhibit. Be able to explain your team's decision to any exhibitor if asked. Always "own" the team's decision. If an exhibitor challenges you on a decision, be prepared to say that the team discussed his concern but concluded otherwise – even if you agree with the exhibitor's comments.

In show competition, judges rely on the accepted evaluation characteristics and their relative weighted values to assess entries. However, judges compare the attributes of entries rather than point scoring them. Most shows are arranged by ADS classification and organized by a show schedule. By using this procedure, blooms are first judged against like cultivars, and then winning entries compete for higher awards in class, section, and show.

Show venues will often include separate divisions for junior, novice, amateur, and open exhibitors. The actual format varies from region to region; some shows group exhibitors by the size of their gardens.

Judges must also exhibit to keep current on horticultural practices, recognize current varieties, and understand the difficulties of exhibiting prime blooms. However, judges should never confuse judging with exhibiting. A judge evaluates what an exhibitor brings to a show, knowing that faults will exist. On the other hand, an exhibitor selects and prepares the best blooms from the garden for a show, knowing that faults exist. Regardless of apparent bloom imperfections, exhibitors should always bring their best blooms and let the judges determine whether they win a prize.

When determining a blue-ribbon winner, judges must know the number of points assigned to each characteristic. This information is on the ADS scorecard. This knowledge will provide them with a better understanding of how severe a certain fault happens to be and will increase their ability to make a fair judgment.

Teams should discuss and evaluate an entry as quickly and in as much depth as possible. They must make an informed decision and be ready to provide the reasons that they have for their choice. Any member of a judging team should be able to provide competent explanations to an exhibitor for the decisions that they have made. Expediency is necessary so the show can be completely judged in a timely manner.

SHOW COMPETITION CONSIDERATIONS FOR JUDGES

Dahlia shows are a comparison of one exhibit to another. Staging is an important consideration and judges should be cognizant of superior staging. See Appendix O. While evaluating each entry, judges are reminded of the following principles:

★ It's important for the new judge to work with experienced mentors who can guide them. The experienced judges will know which characteristics are most important. Picking out a crook in the stem or a nick in a leaf is easy; evaluating form or color faults is much more difficult.

★ Form is the most heavily weighted characteristic; dahlias in competition must be true to type. Blooms exhibiting poor or mixed form should be judged more harshly than those with minor leaf or stem issues. Thus, a bug bite on a ray floret is a more serious fault than a bug bite on a leaf.

★ Color is the second most heavily weighted characteristic when judging dahlias. After form has been evaluated, the judging team should next evaluate color of the entries. Form and color together make up half of the evaluation points; all other characteristics make up the other half.

★ An exhibitor should stage the stem of a single bloom so that it is vertical in the container. With multiple-bloom entries, stems may deviate from a vertical placement. Multiple blooms staged in a single container should not touch. Multiple blooms placed so that they radiate from a point often provide a pleasing display since they appear like many growing plants.

★ For each cultivar evaluated in a show, judges should validate its spelling and classification by checking the latest CHD or composite if they are unfamiliar with a variety.

★ In all instances, judges must assess every entry. On occasion, though, teams may prioritize, identifying entries with obvious faults and selecting superior entries for award consideration. Judges must evaluate all entries, if true to name and correctly entered, on their merits of size, form, or color variations. Cultivars listed in the CHD or classified by a classification committee, correctly placed in a show, cannot be set back from competition because they appear to be incorrectly classified.

★ Many cultivars exist in commerce that have both superior and weak qualities. Therefore, judges should carefully assess the degree of fault of a cultivar and then systematically employ their knowledge of the scorecard valuations to select the best entry. In this process, they must take trueness to form into consideration but must also conscientiously consider each competing cultivar's characteristics before they arrive at their final decision. Whatever the decision, one entry cannot receive added weight for one feature, such as a larger diameter, when the majority of the evaluation characteristics of another entry are superior.

★ Exhibitors spend considerable time growing, transporting, and staging blooms. Keen exhibitors will groom potential show entries in their gardens many weeks before a show to have their entries make the best presentation possible. This is acceptable and considered a good exhibition practice. This grooming includes disbudding, straightening the stem with a mechanical device, using a prop to enhance bloom position, or even severing a spur leaf. Grooming of cultivars also occurs when exhibitors shade the plant to gain color intensity or when they remove the involucres on pompons to produce a bloom with a spherical, ball-like shape. Judges need to be knowledgeable about grooming techniques and should consider poor grooming practices as faults if they detract from the overall presentation of the bloom on the show table. Grooming faults occur when:

- Stubs and side buds remain, or late disbudding is apparent.
- Foliage has been removed and the remaining foliage stubs are above the top of the container.
- Trimmed, removed, or dropped florets are apparent.
- Florets are plucked, leaving obvious gaps in contour.
- Injury or distortion to the stem or stalk is visible because a mechanical device was used incorrectly.
- Apparent damage from the removal of a spur leaf exists.
- Foreign material is used to enhance the bloom position.
- Apparent residue from sprays is evident.
- Pompons do not reflex completely to the stem, leaving exposed involucral bract cuts.

★ Judges may not rearrange ray florets to examine the floral head receptacle or the bloom center nor may they remove the exhibit from the container for any reason. Judges may not restage an entry unless an accident happens. Judges may then carefully restage the entry.

★ Judges should evaluate insect damage by the severity and location. Injury that distorts the ray florets or foliage and is apparent on the face of the bloom is a greater fault than insect injury that is much less conspicuous on the back of the bloom. Dahlia growers have become more cognizant of the dangers to the environment caused by the overuse of chemicals. A minor bug bite to a leaf can be overlooked unless in a tie-breaking situation.

★ Judging from a reasonable distance (arm's length) will help distinguish between faults that are apparent and those that are not. Judge should make sure to employ this strategy as a part of their judging process to determine the best entry.

★ Judges should base their decision of the best entry on its best overall attributes from all bloom characteristics: form, color, substance, foliage, stem, bloom position, uniformity, and proportion of size and depth – not just trueness to type. Team decisions may not be unanimous, quick, or easy, but they will stand since the majority of the team will have made the decision. In this respect, a senior judge's opinion carries no more weight than any other team member. Accredited and candidate judges should certainly listen carefully to the senior judge's reasoning, though, and be willing to change their decision if the senior judge's reasoning makes good sense.

★ If no entry in a class meets the minimum standards for exhibition, judges may award only a red ribbon – or no ribbon – in the

class. The judges should mark each entry tag to indicate the entry was evaluated and not skipped over.

➲ *Appendix C discusses less important characteristics when judging.*

MULTIPLE-BLOOM ENTRIES

Multiple-bloom entries present more of a challenge for both the exhibitor and the judge. Specifics can be found in the local show schedule. In these entries, blooms should never touch one another. Each bloom needs its own space and should be equidistant from the others. Uniformity plays an important role. Judges should assess an entry from whole to part, determining how each characteristic presents itself in all displayed blooms.

Exhibitors may enter multiple-bloom entries in a single container or multiple containers, depending on the show schedule rules. In most shows, triple-bloom entries are common; other multiples may also be listed in the show schedule. Judging multiple-bloom entries is similar to judging single-bloom entries since judges will use the same evaluation characteristics plus uniformity.

Multiple-bloom entries should be uniform in all characteristics. In all views, judges must consider whether the position of the blooms in the entry have balance, proportion, and visual appeal. Lack of uniformity is a serious fault.

Judges must evaluate all blooms and consider each characteristic. Evaluation cannot be weighted toward one bloom in a multiple-bloom entry. Obvious variation of any of the individual evaluation characteristics (color, form, size, depth, substance, stem length, foliage, or bloom position) is a fault in addition to all other faults of the entry.

When evaluating uniformity of open-centered dahlias, judges should adopt a more demanding perspective. The faults for differing qualities of single ray florets will be much greater. Judges should consider the following possible questions: Are the ray florets of equal size and form? Are they equally spaced and arranged in a flat plane? Is the foliage in proper proportion to the bloom? Is the color of the ray florets evenly distributed?

JUDGING UNDISSEMINATED DAHLIAS

Dahlia shows often offer classes for undisseminated dahlias. These are typically single-bloom entries, but some societies may offer multiple-bloom classes. This should not be confused with the ADS Seedling Bench Evaluation which is detailed in the CHD and covered in chapter 4 of this text.

Undisseminated classes give hybridizers an opportunity to have other judges evaluate their new seedling's potential and provides them with another viewpoint. This is especially helpful for the beginning hybridizer. Usually, the society only offers a blue-ribbon award which informs the exhibitor that the seedling should be grown another year. No award indicates to exhibitors that the judging team feels they should no longer enter the seedling competitively.

Judges use the following procedure to judge undisseminated seedlings: they classify each seedling, decide how the seedling compares to what is available in the class, and then award a blue ribbon if the seedling appears to have merit. Judges will employ the same process for both single-bloom and multiple-bloom entries.

EXAMPLE OF A SCORECARD FOR UNDISSEMINATED DAHLIAS

Some dahlia clubs use a structured approach to judging undisseminated dahlias. The following 1-10 scorecard informs an originator of the team's opinion about the seedling's exhibition potential.

The purpose of this scorecard is to provide a consistent method for judging undisseminated dahlias in dahlia shows. The judging team follows this procedure:

- ★ Classify the undisseminated dahlia.
- ★ Determine current competition (refer to current CHD — and current show — as needed).
- ★ Each judge should determine how the undisseminated dahlia would fare against current competition in the open division of that show, first in color class and then at higher levels.
- ★ Each judge should then assign a number from one of the two scales depending on the classification of the dahlia.
- ★ The team should average their scores and write this final score on the entry tag.
- ★ Every dahlia that receives an average score of 4.0 or higher will receive a blue ribbon.
- ★ The dahlia with the highest score in each class will receive a ribbon for best in class. In case of a tie, the judges must decide which

variety possesses the greatest and most outstanding qualities over similar standard varieties.

★ The dahlia with highest overall score will be declared the winner of the award for Best Undisseminated Dahlia in Show. Ties will be broken as above.

If this entry were shown in the open division in a show, this entry would most likely win:

POINTS	SCALE 1 Use for FD, ID, SC, C, and LC Forms	SCALE 2 Use for Other Form Classifications
10	Best in Show	Best in Show
9	Section Champion	Contender for Best in Show
8	Blue Ribbon and Best of Type	Section Champion
7	Blue Ribbon and frequently a Best of Type	Blue Ribbon, contender for Section Champion
6	Blue Ribbon and occasionally a Best of Type	Blue Ribbon and occasionally a Section Champion
5	Blue Ribbon	
4	Sometimes a Red, but most frequently a Blue Ribbon.	
3	Red Ribbon most of the time, occasionally a Blue Ribbon	
2	Red Ribbon	
1	No Award	

1-10 Scorecard, C. Phillip Traff

SPECIAL SECTIONS OR CLASSES

Local show schedules frequently contain special classes. The following are suggested guidelines for many of the special classes typically seen. When judging each of these sections, judges must confirm that the entry conforms to the requirements in the show schedule.

Collections

A collection typically contains a minimum of three individual cultivars. Classes may include blooms of the same form but different colors, blooms of different forms, blooms of cultivars introduced from the local area, or other combinations. These exhibits may be entered in the same or in different containers as specified in the show schedule.

When evaluating collections, judges should pay particular attention to uniformity and color harmony.

People's Choice

Many shows have a separate section designed specifically to be judged by the public. Usually, the cultivars entered in this section possess visual appeal and often consist of bicolor, variegated, and contrasting blends that appeal to the general public. People's Choice winners are not chosen by dahlia judges.

Arrangements

Design elements are the basic visual attributes of which a design is composed. In dahlia shows, the dahlia must dominate the various styles and arrangements. The designs may include line, mass or mass-line, abstract, Oriental, or free form, among others. Whatever the arrangement, it must be the exhibitor's original design. Depending on the show schedule, the exhibitor may or may not be required to grow the dahlias. Exhibitors and

judges should consult the show schedule for specific requirements.

Arrangement entries are often judged by garden club judges who are qualified to evaluate these artistic designs. However, dahlia judges may be called upon to judge an artistic section if specialized judges are not available. While artistic design principles are not part of the dahlia judging curriculum, dahlia judges should familiarize themselves with arrangement design should they be called upon to join the arrangement-judging team.

➲ *See appendix G for a suggested arrangement scorecard.*

Baskets

The main objective of baskets is to exhibit a mass horticultural display of individual blooms, distinctively arranged to show quality and value of each bloom. Judging of baskets is typically assigned to experienced judges; new judges, however, should learn as much as possible by questioning senior judges whenever possible. Check show schedule for specific requirements.

When judging baskets these aspects should be considered:

Basket

- ★ Baskets must have fixed handles unless stated otherwise in the show schedule.
- ★ One-third to one-half of the handle should be visible.
- ★ There should be approximately two to five blooms behind the handle.
- ★ The number of blooms should be in proportion to the size of the basket.
- ★ In close decisions, the condition of the basket may be considered.

Flowers

- ★ Blooms should be arranged so they do not form straight lines.
- ★ No bloom should cover another bloom.
- ★ Blooms should be as uniform as possible in size, form, and color.
- ★ Generally, blooms should be arranged in the shape of the basket.
- ★ Positive and negative space should be about equal.
- ★ Blooms should be checked for substance.
- ★ Any dahlia foliage used should be in good condition.
- ★ No wire or other material may be used to support stems.
- ★ The overall effect should be a dome shape.

Filler

- ★ Filler material such as ferns, huckleberry, or salal should not cover any bloom.
- ★ Filler material should cover staging material.
- ★ Filler material should not detract or overwhelm the blooms.
- ★ Filler material should be free of flowers or berries.
- ★ Baskets should be finished in the back with filler material.

➲ *See appendix H for a suggested basket scorecard.*

Bouquets

Society show schedules will determine whether exhibitors will use only dahlias and natural filler or whether they will be allowed to use other floral varieties and filler. In some shows, exhibitors do not have to grow their own dahlias. The bouquets will be in a single vase and judged from all sides with no visible mechanical distractions such as oasis or wire. Classes may consist of one or mixed cultivars; some show schedules may specify the number of blooms required. Artificial material may not be used.

➲ *See appendix K for a suggested vase or bouquet scorecard.*

Dahlia Bowls

In some locations, an exhibitor may use any size of bowl placed flat or on a pedestal. The exhibit may be viewed from the top and from the side if it is clear glass. Unattached foliage may be used but artificial materials are not allowed. Dahlias ordinarily do not have to be grown by the exhibitor.

In other shows only floating waterlily dahlias are allowed.

Typically, the blooms must be free floating in clear water; specific details are in the club's show schedule. Blooms in this class are typically grown by the exhibitor.

➲ *See appendix N for a suggested bowl scorecard.*
➲ *See appendix J for a suggested floating dahlia scorecard.*

Potted Dahlias

Exhibitors may select their own container, recognizing that it will be judged as a part of the entry. They may support, stake, and/or tie blooms and plants. Most often, classes for potted dahlias are by size or type.

➲ *See appendix M for a suggested potted dahlia scorecard.*

Photography

Exhibitors must submit their own photographs and photos will focus on dahlia(s). Dahlias need not be grown by the exhibitor. The show schedule will dictate the size of photographs, whether they are matted or unmatted, the list of available classes, and the number of entries allowed. Societies will provide an appropriate location for displaying these photographs.

➲ *See appendix L for a suggested photography scorecard.*

CHAPTER 4

JUDGING SEEDLINGS IN THE TRIAL GARDEN AND ON THE SEEDLING BENCH

Seedling Bench Evaluation

Lynn B. Dudley originated the ADS seedling bench evaluation in 1950. The purpose is to determine if a seedling is worthy of being introduced and classified. In this process, if the seedling scores at least 85 on a 100-point scale, the hybridizer names the seedling and it will be included in the next CHD. Appendix B discusses nomenclature control and naming guidelines.

The ADS awards a Lynn B. Dudley medal each year to the highest-scoring seedlings on the bench in these seven categories: AA or A; B; BB; M or MC; BA, MB, or POM; ST, MST, NX, or MNX; WL or MWL.

The Evie Gullikson medal, first awarded in 1999, functions similarly to the Lynn B. Dudley medal except that it is for the highest-scoring open-centered dahlia.

A society show usually has an ADS Seedling Bench Evaluation section; they will follow the established rules printed each year in the CHD. The society must assign only ADS-credentialed accredited and senior judges. The ADS approves and provides the recommended seedling bench evaluation scorecard.

Trial Gardens

The ADS established the first trial garden in 1917. Since then, others have been established throughout the United States and Canada. The purpose of trial gardens is to determine if a seedling is worthy of being introduced and classified by the ADS. If it scores at least 85 on a 100-point scale, it will be included in the next CHD.

The ADS urges all accredited and senior judges to judge in as many trial gardens as possible each year. Derrill W. Hart and Evie Gullikson Medals are awarded each year to the highest-scoring seedlings in the same categories as the seedling bench evaluation.

METHODS OF SEEDLING EVALUATION

To formally evaluate seedlings in trial gardens and on the seedling bench, experienced judges use the ADS scorecard, which can be found on the ADS website and in the CHD.

The candidate dahlia judge will not be called on to judge in ADS seedling bench competition or in a trial garden. They should, however, be aware of the process and should practice judging in these venues under the guidance of a senior judge. After judges achieve accredited status, they may judge the seedling bench and trial gardens.

Judges evaluate trial garden and seedling bench entries relative to perceived perfection as described under Judging Techniques on page 48. They employ an absolute standard of 100 points. They do this by completing an ADS scorecard for each variety judged.

Seedling judges must have a keen understanding of dahlia qualities and have broad experience to accurately evaluate new varieties. They must be able to provide explanations to support their scoring decisions and have a thorough knowledge of dahlias in current competition. Judges who don't grow or exhibit dahlias are therefore not qualified to judge seedlings.

When evaluating seedlings, judges must be open to discussion, while at the same time strong and confident enough to provide a good reason for their decision regardless of venue. Each team member should have the opportunity to voice observed faults or lack of faults.

DETERMINING A CULTIVAR'S FORM

Since a wide range of form variations exists between form classifications, if possible a team of judges should classify each cultivar. If disagreement among these judges prevails, the majority opinion of judges will prevail. Judges evaluating from different regions may disagree when they classify the same named cultivar growing in their respective areas. The form classification represented by the majority of team recommendations from across the country will be the one recorded in the latest CHD.

When determining form, judges must first examine the ray florets near the equator of the bloom, studying what is the regular formation of the ray florets. The next step for judges occurs when they classify the closest, ideal formation of the cultivar. Most often, the form does not adhere strictly to an established form definition and a judge must determine which ideal form definition describes the cultivar most accurately. A dilemma can result. For example, consider a cultivar with a ray floret that is partially involute and pointed at the tip. Judges must decide if the cultivar is an FD, ID, or Ball; if it is regular in arrangement; if it is flat; and if it is fully involute for half its length. No ideal exists. Therefore, judges will determine the degree to which the ray floret in question is involute and will make a calculated decision.

Mixed Form

Although immature, developing ray florets near the center of the bloom and mature ray florets have less of an influence on classification, they should still have a regular appearance similar to the ray florets near the equator of the bloom. When a cultivar has an irregular arrangement consisting of more than one formation, such as SC and ID or ID and LC, a cultivar exhibits mixed form. Mixed form is a major fault.

Any cultivar that competes in a show will exhibit certain faults if judges use the seedling evaluation scorecard to appraise it formally. However, since judges will compare it to other cultivars rather than judging it solely against scorecard standards, they must recognize that it will compete against other cultivars, first in its own class and then against cultivars of other classes as it is considered for higher awards. For it to win in a show, it must exhibit superior qualities that surpass its competition. The scorecard, therefore, is not used when judging in regular show classes. But judges should remember the individual characteristics and the value of each as they compare entries.

Cultivar exhibiting mixed form, which is a major form fault.

The Dahlia Continuum

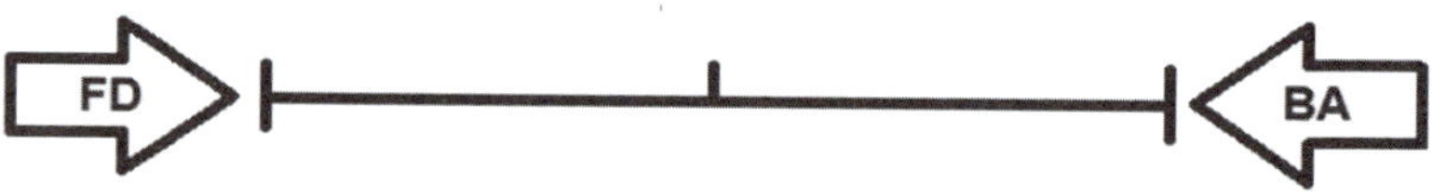

Judges must remember that there is no perfect dahlia and must find the one that comes closest to the stated definition. The example above is for dahlias on the formal decorative to ball continuum. The stated definitions fall at the extreme edges of the continuum in question. In this example, as formal decorative ray florets become more and more involute, cultivars move toward the ball side of the continuum. Those toward the center are "tweeners" and are not of exhibition quality. This is relevant for teams evaluating seedlings as well as for teams judging in the regular horticultural sections of shows.

In cases where a dahlia has mixed form – or such poor form that it cannot win in shows – it should not be classified as a novelty. It is merely a dahlia with bad form.

ADS SEEDLING BENCH EVALUATION

An official ADS seedling bench evaluation can be conducted at any show sponsored by an ADS-affiliated society. Exhibitors must complete an application form for each entry which provides vital information necessary for a cultivar's introduction.

Each entry consists of three blooms that may be staged in one or three containers. At least three judges evaluate each entry; there must be more senior than accredited judges on the team. The judges will use official ADS scorecards. Those entries scoring 85 or higher compete for ADS higher awards. The highest-scoring entry in each show will receive an ADS achievement ribbon. At the conclusion of the show season, the highest three scores shall be used to determine the highest average in each specified class. Winners in each class will receive a Lynn B. Dudley or Evie Gullikson medal.

When judging seedlings in ADS seedling bench evaluation, teams must first collaboratively classify each entry by determining the diameter and depth, stem length, closest formation, and most accurate color and color class. Team members record the results on the ADS entry tag for each variety to be judged. Decisions may not always be unanimous, but the majority opinion will prevail.

After the team members have classified each entry, they then individually "point score" the entries. They judge each characteristic (beginning with form) and record the scores on the ADS scorecards provided. During judging, communication among members of the team is encouraged, especially to point out good or bad characteristics that other team members may have missed. When judging has been completed, the team should compare scores to ensure that one judge's score does

not significantly vary from the others. Judges should talk about the differences. While no changes are mandatory, sometimes a compromise is in order.

When judging on the ADS seedling bench, judges should keep the following in mind:

★ The judge can safely assume that growers entering cultivars in ADS seedling bench evaluation have grown an adequate number of plants, that these cultivars have received above-average care, and that the exhibitor has selected the best three blooms possible. As a result, judges should raise their level of expectations and standards to a higher level than they would in trial garden evaluation where only three plants are available for assessment.

★ Judges must evaluate all three blooms in the entry and score these blooms as one. Judges cannot select only one bloom and disregard the other two. Noticeable differences in any of the evaluation characteristics represent a fault and should be penalized for lack of uniformity.

★ Judges should expect the exhibitor to stage an entry in such a manner that all three blooms can be examined without handling or altering the position of the blooms. The blooms should not be crowded or touch one another.

★ Judges are evaluating the blooms, not the grower's exhibition proficiency. They should therefore evaluate the following items less severely on the seedling bench than they would blooms in open competition:

  - Minor insect damage
  - Minor spray deposits
  - Minor mechanical damage such as bruising of florets.

★ Judges cannot score an entry in which they have an interest.

ADS TRIAL GARDEN EVALUATION

The ADS provides trial gardens across the US and Canada to evaluate how well seedlings grow in various regions. This service furnishes valuable information to originators, introducers, and the public. Seedling growers submit three tubers or plants to the trial garden director who maintains the garden according to

the latest ADS trial garden procedures and rules. Trial garden listings, applicant information, and entry forms are available on the ADS website. Trial garden procedures and rules are listed in the CHD.

Qualified ADS senior and accredited judges evaluate the entries, completing an ADS scorecard for each entry judged. Each trial garden director compiles the scores and submits them to the trial garden chair. The highest scores determine which cultivars will receive a Derrill W. Hart or Evie Gullikson medal.

When judging in an ADS trial garden, judges should keep the following in mind:

- ★ Just as a grower will select and cut blooms for exhibition near sunrise or sundown, judges should avoid scoring in the trial gardens during the hotter portion of the day. If this is unavoidable, then judges should adjust their scoring accordingly.
- ★ To make proper assessments, judges should observe the general growth habits of the plants in the garden, their location, and the husbandry provided under the director's supervision. Are the plants deadheaded, disbudded, disbranched, and tied properly per

the ADS trial garden rules as outlined in the CHD? Excessive or unusual weather conditions should be considered.

★ Judges should determine the size classification of an entry by measuring several blooms. Judges should note if a cultivar's blooms are near a size break and if the number of laterals is not proper for the bloom size. This information could become valuable for those making the final classification determination, since they review reports from several trial garden directors.

★ If a variety is obviously diseased or has some other disorder the entry should receive no score (circle DNS — with the reason — on the scorecard).

★ Often trial garden judging is a solitary experience, but sometimes teams work together. In those cases, the team should collectively determine one classification (size, type, color, stem length) but then may assess the other scorecard characteristics independently.

★ Floriferousness describes plants that produce blooms of high quality and a quantity sufficient for the bloom's size. If a judge does not frequent a trial garden on a regular basis, this characteristic must be determined by the number of blooms, exposed laterals, spent blooms, and buds. Judges who visit a trial garden regularly may provide more accurate assessments about floriferousness.

  - Trial garden judges must consider the number of branches, buds, deadheads, and spent blooms to determine a cultivar's flowering production. Ideally, evaluation of

floriferousness should not only encompass the number of blooms but also their uniform quality.

- If judges who evaluate seedlings late in the season see buds on primary laterals, few blooms, and no evidence of blooms removed, they should fault the seedling for its late-flowering habits.

★ Because judges are evaluating blooms and not horticultural techniques, they should evaluate the following items less severely in a trial garden than they would when judging in regular show competition:

- Minor insect damage unless distortion of the bloom or foliage exists.
- Spray deposits.
- Poor or late disbudding. (All cultivars are supposed to be disbudded to at least the first pair or leaves.)

- Crooked stems and blooms not appearing at the proper angle if an entry has not been tied properly.
- Minor fading on mature blooms located in full sun for the entire day.
- Mechanical damage to blooms; most trial gardens are open to the public.

★ Judges also evaluate the quality of all aspects of each of the three plants entered; judges will evaluate blooms, stem length, size and condition of foliage, plant size and condition when they judge in the trial garden.

★ Trial garden judges do not assess uniformity as a separate characteristic because it is a seedling bench evaluation characteristic.

★ When judging trial garden entries, judges should consider all mature blooms in the evaluation of each characteristic. When there are obvious variances of the measurable attributes of the entry, selected mature blooms with obvious differences will be measured for diameter, depth, and stem length. Judges will determine and record the average size, classification, and stem length.

★ Seedling judges should use the appropriate scorecard designed specifically for fully double or open-centered dahlias.

POINT SCORING IN TRIAL GARDENS AND ON THE ADS SEEDLING BENCH USING THE ADS SCORECARD

Judging seedlings is both a science and an art. Judges will complete a scorecard for each variety, whether judging in a trial garden or in ADS bench evaluation. When first learning to use a scorecard, most new judges tend to score varieties quite low. It is therefore imperative that candidate judges work with a senior judge mentor to practice the fundamentals of using the ADS scorecard both in the trial garden and in ADS bench evaluation.

While evaluating each entry, judges must determine the number of points to deduct for each characteristic. To complicate matters, not all characteristics have the same point value. Form is worth 30 points, color 20, while all others are 10 points each. Thus deducting 2 points (out of 10) for foliage is a higher percentage than deducting 2 points (out of 30) for form. The experienced judge, after evaluating perhaps hundreds of seedlings, will be able to judge a seedling's characteristics and quickly assign points to each. That is why mentorship under a senior judge is so important for the candidate judge: to gain experience, understand the basics of point scoring, and determine if their practice scores are realistic.

When scoring seedlings using the ADS scorecard, a cultivar must receive a score of 85 or higher to pass. While working with a candidate judge, a good mentor will be able to quickly assess most entries and indicate whether a variety should be in the 85-or-higher range. In our example above where 2 points (out of 10) were deducted for foliage, the judge was saying (for foliage), "That is an 80 dahlia." This thought process is a useful tool for each of the scorecard characteristics. For each, the judge needs to evaluate the merits and faults of the entry and then score accordingly. Judges too often deduct too many points from a characteristic with fewer points and too few points from one with many points.

When point scoring, all judges should have a reliable system for determining how many points to deduct for each characteristic on the scorecard. The following percentage calculator

Judging Guidelines				
Characteristic	100%	90%	85%	80%
Form	30	27.0	25.5	24.0
Color	20	18.0	17.0	16.0
Substance	10	9.0	8.5	8.0
Stem	10	9.0	8.5	8.0
Foliage	10	9.0	8.5	8.0
Bloom Position	10	9.0	8.5	8.0
Uniformity Floriferousness	10	9.0	8.5	8.0

is on the reverse of each ADS scorecard and will help the judge determine the proper scores for each characteristic.

To ensure accuracy, judges are encouraged to use decimals rather than fractions on ADS scorecards.

APPENDIX A

CLASSIFICATION

For classification and exhibition purposes, the ADS provides a format by which all dahlia enthusiasts can relate and generalize dahlia formation. By description, a visual image of cultivars becomes apparent. Shows are arranged so that cultivars with similar formations can compete against each other. Although a range of formations will appear in a form class, the range will be much smaller. However, the ADS cannot provide descriptions for all conceivable formations. The important issues are that exhibitors have places to show their blooms and dahlia societies have a system available for organization.

APPENDIX B

NOMENCLATURE CONTROL

Named varieties that receive at least two blue ribbons in a show season or have successfully scored 85 or greater in either seedling bench evaluation or in the trial garden will be added to the CHD. Originators must review the current composite and subsequent CHDs or check the online classification book on the ADS website to verify the name selected for a new cultivar is not already in use. Dahlia name length cannot exceed 22 characters including letters, hyphens, apostrophes, and spaces.

APPENDIX C

LESS IMPORTANT CHARACTERISTICS WHEN JUDGING

When judging dahlias in regular horticultural sections of shows, less experienced judges will be drawn to easy-to-spot defects. Form, color, and the overall presentation of an entry are most important. Minor faults will typically be considered only in tie-breaking situations.

APPENDIX D

RECLASSIFICATION CHANGE REQUESTS AND GUIDELINES

ADS members may suggest changes in classification if they feel a variety is listed incorrectly in the CHD. The procedures to request a change are listed in the CHD under "Cultivar Classification and Change Request Rules and Procedures." Refer to the latest CHD or the most recent June *Bulletin* for the classification change information.

APPENDIX E
DAHLIA SIZES

Dahlias are classified by size and type. The sizes recognized by the ADS are as follows:

Giant (AA) Dahlias............................ Over 10" in diameter
Large (A) Dahlias...................... Over 8" to 10" in diameter
Medium (B) Dahlias.................. Over 6" to 8" in diameter
Small (BB) Dahlias .... Over 4" to 6" in diameter
Miniature (M) Dahlias .... Over 2" to 4" in diameter
Micro (MC) Dahlias ....Up to 2" in diameter
Ball Dahlias.......................................Over 3½" in diameter
Miniature Ball Dahlias........... Over 2" to 3½" in diameter
Pompon Dahlias.................................Up to 2" in diameter
Single Dahlias...................................... Over 2" in diameter
Mignon Single Dahlias ....Up to 2" in diameter

Stellar, Waterlily, Novelty, Peony, Anemone, Collarette, Orchid, and Orchette dahlias are all over 2" in diameter. Micro varieties in those forms are all up to 2" in diameter.

The most recent classification details are listed in the CHD and on the ADS website.

APPENDIX F
ARRANGEMENT SCORECARD TEMPLATE

JUDGING CHARACTERISTIC	SCORE
Design	**40 points**

- ★ The arrangement should have good proportion, balance symmetry, rhythm, a line to follow, or a flowing quality. Some examples are crescent, mass, triangle, line, and Hogarth (S) curve.
- ★ The size of the flowers should be in proportion to the foliage and container. The container should not be the dominant focus.
- ★ There should be a pleasing use of color and contrasts.
- ★ The design should be stable and not appear to be falling over.
- ★ The design should have depth so as not to appear flat or two dimensional.

Cultural Condition	**30 points**

- ★ The blooms should be in good condition, but do not have to meet exhibition standards for form and color. Immature blooms and buds may be used.
- ★ Foliage and other plant material used should be in good condition and have good substance.
- ★ No artificial plant material may be used in the arrangement.

Theme Interpretation **20 points**

- ★ The design should reflect the theme
- ★ The design should adhere to the specifications outlined in the show schedule
- ★ Accessories should complement but not dominate the design
- ★ The main focus of the arrangement should be the dahlias

Distinctiveness **10 points**

- ★ The mechanics of the design should not be conspicuous
- ★ All parts should appear to form a unified, harmonious design
- ★ Originality of design
- ★ Use of unusual materials or materials used in an unusual way

Total Points ________

Entry criteria are set by the individual club and included in the show schedule. These include:

- ★ The size of entries
- ★ Artificial plant material is typically not allowed.
- ★ Typically, no mechanical devices are allowed that require an electrical outlet.
- ★ Plant material need not be grown by the exhibitor, but all exhibits must be arranged by the exhibitor.
- ★ Accessories should be kept to a minimum.
- ★ Dahlias must dominate, although other flowers may be used.

APPENDIX G
BASKET SCORECARD TEMPLATE

<u>QUALITY OF BLOOMS — 50 POINTS</u> <u>SCORE</u>

Form **10 points**

- ★ Major faults: green centers, open centers, not true to type. [-1 to -6] *Note: do not disqualify for green or open centers*
- ★ Minor faults: irregular centers, gaps, thin, immature [-1 to -3]

Color **10 points**

- ★ Major faults: faded, blotchy, streaked, bruised [-1 to -6]
- ★ Minor faults [-1 to -3]

Substance **5 points**

- ★ Faults: ray flowers soft, failing, shriveled [-1 to -5]

Dahlia Stem and Foliage **5 points**

- ★ Stem faults: stem weak, willowy [-1 to -2]
- ★ Foliage faults: if shown, too large, looks diseased, damaged [-2 to -5]

Uniformity **20 points**

- ★ Major faults: different colors in like varieties, mixed and irregular centers [-3 to -10]

QUALITY OF ARRANGEMENT — 50 POINTS SCORE

Proportion and Balance **40 points**

- ★ Blooms too crowded, covering other blooms; no empty space. Not enough blooms for size of basket. Gaps in arrangement [-5 to -15]
- ★ Blooms placed in straight lines or facing unappealing directions [-3 to -10]
- ★ No blooms behind handle, or handle crossing directly in front of blooms, causing poor balance [-1 to -5]
- ★ Arrangement does not follow shape of basket [-1 to -5]
- ★ Unappealing plane: ideal is dome shaped [-1 to -5]

Basket **5 points**

- ★ No fixed handle; ⅓ to ½ of handle not showing; basket in poor condition [-1 to -5]

Filler **5 points**

- ★ Filler covering blooms, not covering staging material, detracting from blooms, not free from berries or flowers; back not finished [-1 to -5]

TOTAL SCORE ______

Note: there is not a fixed score required to achieve a blue or red ribbon; rather scores are to determine rank of exhibits. Entries without competition in a class should be awarded blue ribbons unless there are major faults.

The main objective of a dahlia basket is to be a mass horticultural display of individual blooms, distinctively arranged to show the quality and value of each bloom.

APPENDIX H
CONSIDERATIONS WHEN JUDGING BASKETS

Flowers

- ★ Blooms should be arranged so they do not form straight lines.
- ★ No bloom should cover another.
- ★ Blooms should be as uniform as possible in size, form, and color.
- ★ As a rule, blooms should be arranged in the shape of the basket.
- ★ Positive space (blooms) and negative space (empty) should be about equal.
- ★ Blooms should be checked for substance.
- ★ If dahlia foliage is used, it should be in good condition.
- ★ No wire or other foreign materials may be used to support stems.
- ★ The overall effect should be a dome shape.

Basket

- ★ One-third to one-half of the handle should be showing.
- ★ There should be approximately two to five dahlias behind the handle.
- ★ The number of blooms should be in proportion to the size of the basket.
- ★ In close decisions, the condition of the basket should be considered.
- ★ Baskets should have fixed handles.

Filler

★ Filler material such as ferns, huckleberry, or salal should not cover any bloom.

★ Filler material should cover staging material.

★ Filler material should not detract or overwhelm the blooms.

★ Filler material should be free of flowers or berries.

★ Baskets should be finished in the back with filler material.

Common Mistakes

★ Lack of uniformity in color/bloom size/maturity or blooms.

★ Overcrowding in proportion to basket.

★ Too few blooms.

★ Baskets too much like arrangements.

★ Flat face to the basket.

★ Lack of understanding in use of scorecard.

APPENDIX I

FLOATING DAHLIA SCORECARD TEMPLATE

JUDGING CHARACTERISTIC	SCORE
Quality of Blooms	**60 points**
★ Waterlily dahlias preferred; often required ★ Stem length less than 2 inches	
Bloom/Bowl Proportion and Harmony	**25 points**
★ Bloom should have freedom of movement in bowl ★ Petals should not touch side of bowl ★ Bowl should be wider than it is tall ★ Water in bowl should be clear ★ Other accessories are not allowed, unless specified in the show schedule ★ Greenery may be allowed	
Distinctiveness	**15 points**

APPENDIX J

VASE SCORECARD TEMPLATE

JUDGING CHARACTERISTIC	SCORE
Quality of Blooms	**25 points**
★ Fresh and undamaged; blown centers are a fault but not a disqualification	
Balance and Proportion	**25 points**
★ Should not appear top-heavy, but should have an even distribution of various types and cultivars	
Spacing of Blooms	**20 points**
★ Blooms crowded and distorted by neighboring blooms and/or greenery is a fault.	
Color Harmony	**20 points**
Distinctiveness	**10 points**

APPENDIX K
PHOTOGRAPHY SCORECARD TEMPLATE

JUDGING CHARACTERISTIC	SCORE
Presentation/Matting	**20 points**
Subject/Technical Quality or Standard	**40 points**
Style/Wow Factor (emotional impact)	**40 points**

Subject/Technical Quality or Standard — **40 points**

- ★ True to class – 10 points
- ★ Quality of subject – 10 points
- ★ Technical quality of photography – 10 points
- ★ Minimal distraction diminishing subject impact – 10 points

Style/Wow Factor (emotional impact) — **40 points**

- ★ Proportion of subject to frame – 10 points
- ★ Position of subject in frame – 10 points
- ★ Angle of shot – 10 points
- ★ Creative thinking or technique used to achieve wow factor – 10 points

APPENDIX L

POTTED DAHLIA SCORECARD TEMPLATE

JUDGING CHARACTERISTIC	SCORE
Cultural Perfection	**35 points**
Floriferousness	**30 points**
Foliage	**20 points**
Condition and suitability of container	**15 points**

APPENDIX M

BOWL SCORECARD TEMPLATE

JUDGING CHARACTERISTIC	SCORE
Technical aspects and condition ★ Clean, fresh, properly conditioned blooms and foliage	**40**
Distinction and creativity	**20**
Staging and artistic appearance	**20**
Condition and suitability of container	**10**
Size of blooms in proportion to container	**10**

APPENDIX N
SAMPLE SHOW JUDGING PROCEDURE

Before starting make sure all entries are in the proper order and in their proper classes. It may be helpful to separate blooms from a crowded show table by employing a separate judging table or cart. This allows the team to focus on the blooms being judged without the distraction of the other blooms.

The severity of faults is proportional to the distance from which they are apparent. A fault apparent from six feet away is more severe than one only apparent from arm's length. Similar faults on the face and front of exhibits are more severe than those same faults on the sides or back. Therefore, start the judging process from at least arm's length or more.

All judges should take turns standing directly in front of each entry to evaluate all characteristics. Determine if it is on center with equal amounts of the bloom on each side of the stem.

The team should discuss any other observations that are apparent, especially balance and proportion.

Now evaluate the entry from above. Often this is facilitated by placing the entry on the floor. Centers should be ranked for symmetry and maturity. Bloom contour, symmetry and gaps should be discussed along with the overall color uniformity.

The entry can be returned to the table and viewed from all sides. Observations should be discussed including the depth to diameter ratio. Substance and maturity should also be discussed. The bottom of the blooms should also be viewed.

Finally return the entries to their forward-facing positions and step back one more time. As a team, summarize and rank all the attributes and faults that have been discussed. Take time to arrive at a consensus decision to rank the entries so that the entire team feels they have contributed and are comfortable with the decision. If a consensus cannot be reached even after thorough discussion, a vote must be taken.

All team members should be able to reconstruct the decision-making process that led to the team's conclusion when asked.

APPENDIX O

STAGING DAHLIAS FOR SHOWS

Staging is the art of presenting a dahlia bloom by an exhibitor in the most eye-pleasing manner possible. Exhibitors stage blooms to maximize the balance and proportion of an exhibit. This includes deciding how much foliage and stalk to include to draw the viewer's eye to the bloom. For example, an extra set of leaves may enhance or detract from this goal.

- ★ Entries when placed in exhibition should allow both judges and the public to view the blooms at an appropriate height.
- ★ While staging of blooms is not an evaluation characteristic, it may influence the final decision of a judge. It is often the staging decisions that an exhibitor makes to maximize the beauty of an exhibit that is the difference between entries.
- ★ The ADS requires that a cultivar must have at least one pair of attached leaves when exhibited. The stem and stalk of an entry should be straight and vertical.
- ★ For larger blooms, the container and filler (or wedge) must adequately support the bloom(s). A well-staged entry will not have filler extending above the staging container top and wedges will extend only slightly above the container.
- ★ Ideally, the diameter of the bloom, the length of the stem, and the size of the foliage should reveal a pleasing balance and proportion. A small bloom on a long stem with small foliage does not possess good proportion, nor does a large bloom on a short stem with large foliage. Regardless of bloom size, judges will assess a fault when foliage is too large for the

size of the bloom or nearly touches it. The overall spread of the foliage should at least be as large as the diameter of the bloom.

- ★ In special classes such as floating dahlias, baskets, photography, and arrangements, exhibitors do not need to include leaves unless the local show schedule requires them.
- ★ All blooms entered in show competition will be properly disbudded; the exhibitor will disbud leaving little evidence.
- ★ Cultivars may have a spur leaf of varying size between the bloom and first pair of true leaves. A small spur leaf located approximately one-third of the distance from the bloom may appear as aesthetically pleasing, especially on longer stems. A larger spur leaf will create an unbalanced symmetry and is a fault. Some exhibitors may choose to remove a spur leaf early in its growth. A judge should not fault this removal if it is undetectable at a reasonable distance (arm's length). If the removal of a spur leaf is obvious, judges will assess a penalty.
- ★ An ideal pair of leaves exhibits balance on a horizontal plane and a uniform shape. They will connect directly at a node. On occasion, a set of leaves appears and does not attach directly at the node, growing up to an inch apart. This condition is a minor fault. Judges should understand that these leaves are not considered as unpaired. When leaves stair step along the stem length, they are unpaired, and it is a fault. Properly conditioned leaves will be firm, clean, crisp, uniformly colored, and free of foreign substance, mildew, and insect or mechanical damage.

- ★ Judges should not set back an entry from further competition based on one observable minor fault. Faults are always a matter of degree. For instance, insect injury on the visible face of an entry should be determined as a greater fault than a somewhat inconspicuous injury on the reverse. The judge must evaluate the entire entry, comparing it to its competition before making such a determination.
- ★ The stem should be round, smooth, and in proper proportion, not tapering excessively for its entire length. Judges may fault stems that are too willowy or too thick and club-like. The removal of patina from a stem may enhance the overall appearance of the entry. If the patina or natural bloom appears blotchy, it is a minor fault.
- ★ A judge may fault an entry when foliage touches the container, is submerged into it, or when stubs from removed foliage appear above the container.
- ★ The presence of spray deposit or minor insect damage may be a fault but should not be a reason for setting back an entry; judges must consider the overall condition of the exhibit.

APPENDIX P

ADS AWARDS

The Derrill W. Hart medal

Awarded to the cultivar that has the highest average trial garden score in the accredited trial gardens located across the United States and Canada during the previous year as an AA/A; B; BB; M; Ball/Miniature Ball/ Pompon; ST/NX; and Waterlily. The highest three scores will be used to determine such highest average. In case of ties, duplicate awards will be presented.

The Lynn B. Dudley medal

Awarded to the cultivar that has the highest average seedling bench evaluation score as an AA/A; B; BB; M; Ball/Miniature Ball/ Pompon; ST/NX; and Waterlily). The highest three scores awarded at ADS affiliated shows in the United States and Canada will be used to determine such highest average. In case of ties, duplicate awards will be presented.

The Evie Gullikson medal

Awarded to the cultivar that has the highest average Seedling Bench Evaluation score and the cultivar with the highest average Trial Garden score as an open center or disc center. In case of ties, duplicate awards will be presented.

The Stanley Johnson Medal

Awarded to the cultivar, which has not previously won a Stanley Johnson Award, with the greatest number of higher than blue ribbon awards that were won at the shows held during the previous year by ADS member societies in the United States and Canada. In case of ties, duplicate awards will be presented.

The Les Connell Medal

Awarded to the cultivar that was introduced five years ago with the most awards (blue ribbons and higher) in the ensuing five-year period. For example, at the end of the 2025 growing season the Les Connell Medal will be awarded to the dahlia cultivar introduced in 2021 which has the most awards (blue ribbons and higher) in the five-year period from 2021 to 2025. Due to the time restriction for this award, it can only be won once. In case of ties, duplicate awards will be presented.

GLOSSARY

ADS. American Dahlia Society.

ADS Accreditation Chair. The committee chair appointed by the ADS president to administer the accreditation of judges.

ADS Seedling Evaluation Scorecard. A formal document used when judging cultivars in a trial garden or in ADS seedling bench evaluation.

Bearded

Accredited Judge. A judge who has satisfactorily completed certain requirements of training and judging and continues to grow and exhibit dahlias.

Achromatic. Colors that lack hue and chroma (neutral colors: black, white, and gray).

Arm's Length. Approximately three feet.

Asymmetrical Balance. Approximate equal visual weight composed of different elements on each side of a vertical axis. Balance without symmetry.

Basket. A container with a handle; may be composed of various materials. May be viewed from the front or all sides depending on the show schedule.

Blown

Bearded. A bloom with longer ray florets below the centerline of the flower head creating an asymmetrical appearance. Fault by degree.

Bicolor. Two or more distinct, clear, and sharply separated colors on the face of a ray floret.

Bleed(ing). Two or more colors intermingling in bicolor dahlias rather than a clear separation of colors. A minor or major fault, depending on amount of bleeding.

Blend. Two or more merging colors on either the face or reverse of the ray floret OR two or more contrasting colors on either the face or reverse of the ray floret.

Bloom. (1) The complete inflorescence in a composite flower. (2) The natural gray, powdery substance found on the stem of dahlias. (3) Patina.

Blown. Any fully double dahlia that has over-matured to expose the disc flowers. A major fault (eliminate).

Blush. A light, uniformly tinged coloration on florets.

Bract. A modified leaf; three types of bracts are associated with a dahlia bloom or stem. (1) Palea bracts – the scale-like structures located at the base of each ray floret. (2) Involucral bracts – these make up the involucre that encloses the bloom while in bud and attach to the receptacle (3) Stem bracts (aka spur leaf) – the leaf-like structures that appear between the bloom and the first pair of leaves. A minor fault when palea bracts are conspicuous from three feet;

Examples of petal formation, top to bottom: Semi-Cactus, Cactus, Incurved Cactus, Laciniated.

the fault is greater when the bracts are more apparent.

Bruised. Flower parts, stem, or foliage damaged by weather, grooming, or handling. A minor or major fault depending on competition and amount of damage.

Bud. A small axillary or terminal protuberance containing the rudimentary inflorescence; most often an axillary bud appears on each side of the terminal bud.

Build. The combined parts of a fully double bloom that give it composition and structure including a symmetrical contour viewed from all positions and centered on the receptacle. For most forms, the ideal build will have a high floret count.

Bull Center, Bull Nose, Bull Head. A bloom with a large or malformed center, out of proportion, or with a solid green center.

CHD. See *Classification and Handbook of Dahlias.*

Candidate Judge. A beginning judge who has completed the necessary coursework. Requirements are on the ADS website.

Coarse

Central Rays. The short or immature ray florets comprising the central portion of the fully double flowers when at their prime stage.

Character Leaf. See *spur leaf.*

Chromosome. A nuclear body in a cell having definite structure and bearing, determiners of heredity, or genes.

Chroma. A dimension of color. The strength or weakness of a hue. The degree of intensity or grayness, the purity of color.

Class. A grouping of dahlias of the same size, form and/or color.

Classification and Handbook of Dahlias. An annual ADS publication listing dahlias currently being exhibited as well as classification rules, form definitions, color definitions, and other valuable information.

Coarse. Ray florets too large, out of proportion to the size of the bloom, unevenly arranged, and/or with a wide inconsistent spacing between the ray florets. Often in varieties with low ray floret count. Fault by degree.

Collection. A multiple-bloom entry of dahlias, the size and purpose as specified in a show schedule.

Color. The quality of a substance with respect to light reflected, usually determined visually by measurement of hue, saturation, and brightness of the reflected light.

Color Chroma. A dimension of color; the strength or weakness of a hue. The degree of intensity or grayness; the purity of color.

Color Harmony. A pleasing, compatible combination of colors.

Color Hue. The name of a color.

Color Value. The degree of lightness and darkness of a color.

Compositae. Belonging to a family of plants characterized by each floret producing a seed, being floriferous, leaves appearing alternate and opposite, containing all ray florets in the floral head surrounded by an involucre. Compositae examples include the daisy, dandelion, aster, sunflower, and dahlia among 1,600 to 1,700 other genera.

Composite Head. An inflorescence composed of many tightly packed, small ray and disc flowers.

Composite Listing of Dahlias. An ADS publication listing every dahlia included in the CHD since 1976. Published approximately every five years.

Condition. The physical state of plant material, ideally clean, crisp, fresh, free of wilt, injury, or disease.

Conditioning. Preparation of plant material after cutting for exhibition.

Contour. The outline/silhouette of the face or side extremities of the dahlia.

Crotched. Bloom too close to the first set of leaves.

Cultivar. A variety of plant produced by hybridization and maintained by vegetative propagation.

Cultural perfection. A cultivar exhibited at peak maturity, free of blemishes, attaining its full potential.

Demeanor. The way in which judges behave or conduct themselves.

Development. The state of maturity of the bloom. Fault by degree from ideal peak of development.

Disbranch. To remove secondary branches at the leaf nodes to limit the number of blooms on a plant.

Disbud. The removal of two lateral (axillary) buds early in the plant's growth to produce a superior terminal bud.

Disc: A more or less flat, circular group of tubular florets fully visible in open-centered dahlias.

Disc Flower. Small tubular ray florets that make up the central part of a composite head, each with pistil and stamen.

Disc-centered bloom. See open-centered bloom.

Disqualify. An entry eliminated in the initial evaluation of a class due to an infraction of show rules or obvious disease.

Disseminated: A variety that is available to others, typically through sales channels. See undisseminated.

Disc Flower

Distinction. A judging characteristic previously used on ADS scorecards. Used to describe cultivars that stand out from competition. Dropped from the ADS scorecard after 2021.

Double. A bloom with multiple rows of ray florets; the disc florets are immature and completely covered by the central ray florets when the bloom is at its peak of development.

Eliminate. All entries of a class are evaluated, and the entries with the greater faults are considered not worthy of an award and set back from further consideration.

Eye zone. A uniform band of a contrasting color(s), less than one-third the length of ray florets that surround the open-centered disc flowers.

Fimbriated. See laciniated.

Flicked floret. A ray floret that is twisted, misshapen, and impairs the outline of a bloom.

Floral Art. See flower arrangement.

Floret. One of the small flowers that collectively comprise the flower head. See Compositae.

Floriferous. Vigorous plants bearing many flowers.

Green Center

Flower Arrangement. A decorative arrangement of flowers and other natural materials incorporating the principles of balance, proportion, rhythm, contrast, and unity. Flower arrangements in ADS-affiliated shows may not include artificial plant material.

Fluted. Ray florets that are grooved. A fault if not uniform.

Fully Double Flower. Flower heads with multiple rows of ray florets; the disc florets are immature and completely covered by the central rays when a bloom is at the prime stage.

Gappy, Gaps. A condition in the contour of the bloom where the ray flowers have uneven spacing in a part of the floral head. Fault by degree.

Green Center. Ray flowers making up the center are still undeveloped; the center appears hard and green. Major fault.

Head (also known as floral head). A rounded, full inflorescence.

Honorary Senior Judge. A senior judge who has served many years but, for reasons as determined by the club/conference evaluation chair, is no longer fully able to fulfill the requirements for maintaining their judging status.

Hue. The name of a color.

Ideal. Perfect. The highest conception of standard for imitation.

Immature. A bloom not fully developed.

Incurved. A ray floret curves at the tip and along its par-

Involucre

tial length toward a bloom's center.

Inflorescence. A receptacle holding clusters of flowers; a bloom.

Involucre. A circle or collection of bracts that enclose a developing bud, attached to the receptacle of the floral (composite) head. Fault if involucre has been cut and is observable.

Involute. Margins of the ray floret roll inward along their longitudinal axis (length). When fully involute the margins touch or overlap so only the reverse of the floret is visible, appearing quill like. Opposite of revolute.

Laciniated. The tip of the ray floret exhibits single or multiple splits, with adequate depth; tips either involute or revolute and should twist in an irregular manner to present a fringed/frilled effect.

Lateral. The growth at the leaf node of the stalk, secondary laterals grow from the leaf nodes of the primary laterals.

Margin. The edge of the ray floret, sometimes marked with a narrow contrasting color.

Marginal Rays. Fully developed ray florets that establish and determine the maximum diameter of a bloom; mature florets rather than immature central ray florets.

Majority of Ray Florets. Located in the general equatorial area of the bloom, the ideal location to determine form classification.

Micro. A diminutive dahlia up to 2 inches in diameter.

Mignon. Single dahlias up to 2 inches in diameter; plants are usually not tall.

Notched Florets

Mixed form. A bloom exhibiting ray florets of more than one type. A serious fault. See true to form.

Neck. Junction between the bloom and the stem.

Negative Space. Empty space. The absence of ray florets in a bloom or the absence of blooms in a basket or multiple-bloom entry.

Node. Slightly enlarged portion of the stem where foliage and secondary laterals originate.

Notched Florets. Florets that are slightly split at the tip. Not true to form.

Open-centered bloom. A bloom with one or more rows of ray florets that surround a visible disc. Examples include peony, collarette, single and several other forms.

Pair of Leaves. Two leaves located at the node that appear directly across from each other. See set of leaves.

Patina. The natural gray, powdery substance found on the stem of dahlias.

Penalize. The identification and assessment of faults when evaluating entries.

Petal. See ray floret.

Picotee

Petaloid. Additional floral parts with a form similar and usually smaller than the adjacent ray floret; sometimes misshapen or twisted. Required on collarette and orchette dahlias. Major fault on all other types.

Picotee: A fine line of contrasting color found on the margin of ray florets.

Pig Ear. See spur leaf.

Positive Space. Space taken up by ray florets in a bloom or blooms in a basket or multiple-bloom entry.

Preferred. More desirable trait than another. For example, on collarette dahlias, rounded petals are preferred over pointed petals.

Proportion. The perceived balance of the relationship

between the bloom, stem, and foliage.

Plucked Bloom. Ray floret(s) removed creating voids, gaps.

Ray Floret. A single petal of a bloom that contains the normal flower parts except for a stamen.

Receptacle. Point of attachment of the stem and floral head.

Recurved

Recurved. A ray floret curves away from the center along its length toward the stem.

Reflexed. A straight ray floret that extends and drops toward the stem.

Revolute. Margins of the ray florets roll back along their longitudinal axis (length). When fully revolute the margins touch or overlap. Opposite of involute.

Scorecard. See ADS Seedling Evaluation Scorecard.

Seedling. An undisseminated cultivar grown from seed, often used as identification until named and entered into commerce.

Self-Colored Bloom. A bloom with one color.

Senior Judge. Designation used by the ADS for the highest judging level; a judge who has satisfactorily completed additional requirements of training including trial garden judging, seedling bench evaluation, and growing and exhibiting dahlias. Often serve as mentors for candidate and accredited judges.

Set Back. Eliminated from further consideration.

Set of Leaves. Leaves that do not attach directly at the node, growing up to an inch apart. Fault by degree if more than an inch apart. See pair of leaves.

Shade. A hue mixed with black produces a shade.

Shallow Bloom. A bloom with inadequate depth.

Sport. A mutation, sudden variation or change of color, form, size, or other characteristic of the plant or plant part.

Spur Leaf

Spur Leaf. A single leaf located between the receptacle of the flower and the first set or pair of true leaves.

Staghorn ray florets. Two adjacent ray floret tips that turn in opposite directions causing a gap. Most often appears in incurved formations.

Stem/Stalk. The stem is the support of the flower from the receptacle of the flower to the first pair or set of true leaves; below this point is the stalk.

Substance: A perception of the freshness of plant material. In dahlias, ray florets and leaves should be firm, crisp, springy, fresh, and hydrated. See turgid.

Sunken or Hollow Center. Center of the bloom below the general outline leaving a depression.

Symmetry. Equal balance of weight or material on both sides of a central axis.

Tint. A hue mixed with white gives tint.

Tone. A hue mixed with gray produces a tone.

Tri-leaf. Three leaves attached at the node. Leaves are not alternate and opposite.

True to Form, True to Type. Meeting the definition for a dahlia form or type. See mixed form.

Turgid. Fully filled with water. A turgid dahlia has good substance.

Undisseminated. A variety that is not available to the public. Typically, an unreleased seedling. See disseminated.

Unpaired leaf. See spur leaf.

Staghorn Ray Florets

Wolf Petal

Variegated. Two or more distinct colors on the face or reverse of the ray florets, the variegating color arranged in dots, flecks, splashes, stripes, or narrow lines, which contrast with the basic color(s).

Willowy. A thin or weak stem that appears too delicate for the size of the bloom.

Wolf Petal. A floret that possesses a different coloration from the rest of the bloom.

INDEX

NOTES

NOTES